Recent Advances in Bacterial Biofilm Studies - Formation, Regulation, and Eradication in Human Infections

Edited by Liang Wang, Bing Gu, Li Zhang and Zuobin Zhu

Published in London, United Kingdom

Recent Advances in Bacterial Biofilm Studies - Formation, Regulation, and Eradication in Human Infections
http://dx.doi.org/10.5772/intechopen.108089
Edited by Liang Wang, Bing Gu, Li Zhang and Zuobin Zhu

Contributors
Jorge Manuel Silva-Jara, Carlos Arnulfo Velázquez, Dulce María Romero-García, Jazmín Guadalupe Silva-Galindo, Alejandra Monserrat Castorena-Sánchez, Ismael García-Vera, Gabriela Hinojosa-Ventura, Zahra Sedarat, Andrew W. Taylor-Robinson, Zeuko'o Menkem Elisabeth, Chin Erick Ngehdzeka, Salma Kloula Ben Ghorbel, Sana Dhaouadi, Sana Bouzenbila, Ameur Cherif, Ramzi Boubaker Elandoulsi, Sandile Phinda Songca, Liang Wang, Muhammad Usman, Huan Yang, Jun-Jiao Wang, Jia-Wei Tang, Li-Yan Zhang, Bing Gu, Matthew Terzungwe Tion, Kenneth Ikejiofor Ogbu, Felix Kundu Shima, Muhammad Akbar Shahid, Ali Saeed, Sadeeq Ur Rahman, Mian Muhammad Salam, Sheraz Ahmad Bhatti, Muhammad Mudaseer Nazir, Muhammad Nauman Zahid

First published in London, United Kingdom, 2024 by IntechOpen
IntechOpen is the global imprint of INTECHOPEN LIMITED, registered in England and Wales, registration number: 11086078, 5 Princes Gate Court, London, SW7 2QJ, United Kingdom

British Library Cataloguing-in-Publication Data
A catalogue record for this book is available from the British Library

Additional hard and PDF copies can be obtained from orders@intechopen.com

Recent Advances in Bacterial Biofilm Studies - Formation, Regulation, and Eradication in Human Infections
Edited by Liang Wang, Bing Gu, Li Zhang and Zuobin Zhu
p. cm.
Print ISBN 978-1-80356-708-2
Online ISBN 978-1-80356-709-9
eBook (PDF) ISBN 978-1-80356-716-7

For EU product safety concerns:
IN TECH d.o.o., Prolaz Marije Krucifikse Kozulić 3, 51000 Rijeka, Croatia,
info@intechopen.com or visit our website at intechopen.com.

Meet the editors

Prof. Liang Wang is a group leader and principal investigator at Guangdong Provincial People's Hospital (Guangdong Academy of Medical Sciences), Southern Medical University, China. He also works as a research fellow at the University of Queensland and the University of Western Australia. He is an adjunct professor at the School of Medical and Health Sciences, Edith Cowan University, Australia. His current research interests are computational and experimental bacteriology, focusing on bacterial evolution, metabolism, and rapid diagnosis. Prof. Wang serves as an associate editor for *Frontiers in Microbiology* and is an editorial board member of *BMC Microbiology*, *PeerJ*, and *Translational Metabolic Syndrome Research*. To date, Prof. Wang has published more than 100 peer-reviewed papers in international journals.

Professor Bing Gu is the director of the Department of Laboratory Medicine at Guangdong Provincial People's Hospital (Guangdong Academy of Medical Sciences), Southern Medical University, China. He is a visiting scholar at UCLA and Purdue University, USA. Dr. Gu has published more than 160 peer-reviewed papers. He is Vice Chairman of the Youth Committee of the Chinese Society of Laboratory Medicine of the Chinese Medical Association and Vice President of the Society of Laboratory Medicine of the China Association of Medical Equipment.

Li Zhang received her MBBS from Fudan University, China, and her Ph.D. from the University of Cambridge, UK. She was a clinician at the China-Japan Friendship Hospital, in Beijing before her Ph.D. study. After completing her Ph.D., she worked as a postdoctoral fellow at the National University of Singapore before joining the University of New South Wales (UNSW). Associate Prof. Li Zhang's group investigates bacterial species that cause or prevent inflammatory diseases and cancers of the gastrointestinal tract. They study bacterial genomes, bacterial virulence factors, interactions of bacterial pathogens and gut microbiome with the host immune system, and novel methods to modify gut and oral microbiome and to enhance antibiotic sensitivity.

Zuobin Zhu is an associate professor at Xuzhou Medical University, China. Since 2016, he has been researching microbe–host interaction and the genetic mechanisms of infertility. He has authored more than thirty papers in refereed international journals on drug resistance mechanisms of pathogenic bacteria, the relationship between intestinal microbes and host health, and the genetic mechanism of complex diseases, especially infertility. He has edited or co-edited two books. Dr. Zhu is a member of the Microbial Genetics Committee of the Genetic Society of Jiangsu Province, a member of the Male Reproduction Branch of the Jiangsu Maternal and Child Health Care Association, and appointed head of the Xuzhou Engineering Research Center of Medical Genetics and Transformation.

Contents

Preface

Infectious diseases are a global public health priority due to the heavy burden they place on the healthcare system and the high costs of their management [1]. These diseases are caused by infection by microorganisms such as bacteria, viruses, fungi, and parasites. Therefore, it is important to prevent the spread of these microorganisms to reduce the transmission of infections from person to person. Among all the microorganisms, bacteria cause many common infections, such as pneumonia, sepsis, gonorrhea, and so on; however, not all bacteria are harmful. Controlling pathogenic bacterial infection is critical in preventing and reducing global human infection rates and mortality. Antibiotics are a group of low-molecular-weight active antibacterial substances used to target bacterial infections for more than 70 years [2]. However, the abuse of antibiotics has led to the rapid rise of multidrug-resistant pathogenic bacteria, causing a paradigm shift in the relationships between antibiotics and bacterial–human relations [3]. In addition, bacteria can form biofilms consisting of surface-associated microbial cells with exopolysaccharides, extracellular DNA, proteins, and amyloidogenic proteins enclosed in an extracellular polymeric substance matrix [4, 5]. Various studies confirm that bacterial biofilms protect bacteria from harsh conditions and play critical roles in antibiotic resistance [6]. Although the role of biofilms in antibacterial resistance is complex, it is found that many human infections are biofilm-mediated, and biofilms can significantly drive resistance. Therefore, it is important to understand biofilm formation and regulation to facilitate biofilms' rapid and effective eradication, which will further increase the prevention and control of bacterial infections in human beings. Over the past decade, a large amount of work has been conducted and reported in medical literature. However, due to the rapid progress in bacterial biofilms, there is a need to compile recent important works in a single volume. As such, we have published this book, *Recent Advances in Bacterial Biofilm Studies – Formation, Regulation, and Eradication in Human Infections.* This edited book contains chapters written by international teams of basic and clinical researchers. These chapters provide novel insights and advanced knowledge of bacterial biofilms in clinical settings for life science researchers, clinical researchers and doctors, and other interested readers. They aim to facilitate the understanding of bacterial biofilms and strengthen the importance of novel methods and technologies in the treatment of biofilm-related human infections. With the development of prevention, inhibition, and eradication methods of bacterial biofilms, the mortality rate of chronic and fatal bacterial infections is expected to be greatly reduced in the future.

Many people have contributed to this book. First, I would like to thank my Publishing Process Manager, Ms. Nina Miocevic, for her patience and expertise in assisting us throughout the publication process. I would also like to thank the Commissioning Editor, Ms. Sandra Bakic, for providing us with the opportunity to edit the book.

We also acknowledge the contributions of the production and typesetting teams for their professional services. We hope you enjoy reading this book as much as we have enjoyed writing it.

Liang Wang
Group Leader, Principal Investigator,
Department of Laboratory Medicine,
Guangdong Provincial People's Hospital,
Guangdong Academy of Medical Sciences,
Southern Medical University,
Guangzhou, China

Bing Gu
Laboratory Medicine,
Guangdong Provincial People's Hospital (Guangdong Academy of Medical Sciences),
Southern Medical University,
Guangzhou, China

Li Zhang
University of New South Wales (UNSW),
Kensington, Australia

Zuobin Zhu
Associate Professor,
Xuzhou Medical University,
Xuzhou, China

References

[1] Ikuta KS, Swetschinski LR, Robles Aguilar G, et al. Global mortality associated with 33 bacterial pathogens in 2019: A systematic analysis for the Global Burden of Disease Study 2019. The Lancet. 2022;**400**:2221-2248

[2] Uddin TM, Chakraborty AJ, Khusro A, et al. Antibiotic resistance in microbes: History, mechanisms, therapeutic strategies and future prospects. Journal of Infection and Public Health. 2021;**14**:1750-1766

[3] Van Giau V, An SSA, Hulme J. Recent advances in the treatment of pathogenic infections using antibiotics and nano-drug delivery vehicles. Drug Design, Development and Therapy. 2019;**13**:327-343

[4] Donlan RM. Biofilms: Microbial life on surfaces. Emerging Infectious Diseases. 2002;**8**:881-890

[5] Sharma D, Misba L, Khan AU. Antibiotics versus biofilm: an emerging battleground in microbial communities. Antimicrobial Resistance & Infection Control. 2019;**8**

[6] Abebe GM. The role of bacterial biofilm in antibiotic resistance and food contamination. International Journal of Microbiology. 2020;**2020**:1-10

Section 1

Introduction

Chapter 1

Introductory Chapter: Bacterial Biofilms in Human Infection – An Urgent Problem That Needs to Be Solved

Liang Wang and Bing Gu

1. Introduction

Infectious diseases pose continuous and increasing risks to human health and welfare along with the development of human society [1]. Even nowadays, in the twenty-first century when individuals, communities, and hospitals have easy access to effective disinfectants, abundant antibiotics, and advanced medical technologies, infectious disease outbreaks are still able to cause severe consequences on lives and livelihoods all over the world [2]. In fact, according to a recent comprehensive demographic analysis for the Global Burden of Disease Study, around seven million people died of infectious diseases in 2019, representing approximately 12% of all deaths globally [3]. As one of the five major infectious agents, that is, viruses, bacteria, fungi, protozoa, and helminths [4], bacteria have a significant impact on public health since their infections can occur at any part of human body and can be transmitted to human beings *via* many apparent and hidden routes, making their control and prevention extremely difficult [5]. Bacterial infections have caused many devastating pandemics in human history such as the Black Death caused by *Yersinia pestis*, Cholera caused by *Vibrio cholerae*, and so forth, leading to tens of millions deaths all over the world [6]. In addition, there are many other bacterial pathogens that are able to cause infectious diseases with severe consequences like leprosy (*Mycobacterium leprae*), tuberculosis (*Mycobacterium tuberculosis*), anthrax (*Bacillus anthracis*), syphilis (*Treponema pallidum*), and so on, making clinical treatment of bacterial infection difficult and complex. Since the discovery of penicillin in 1928 and other antibiotics after that, due to the high effectiveness of antibiotics in bacterial infection control, it was once thought that antibiotics were one of the most successful chemotherapies in the history of medicine [7]. However, with the arm race between bacterial evolution and the development of antibiotics, drug-resistant and multidrug-resistant bacterial pathogens keep emerging, leading to the emergence of hard-to-treat multiple antibiotic-resistant infections and the failure of last-line antibiotics [7]. In clinical settings, formation of bacterial biofilms could significantly enhance antibiotic resistance from 10- to 1000-fold increment when compared to similar bacteria living in a planktonic state [8]. In addition, bacterial biofilms are closely associated with chronic infections like pneumoniae in cystic fibrosis patients and are extremely difficult to eradicate once formed [9]. Therefore, it is important to understand bacterial biofilm formations and regulations at molecular level so as to facilitate the development of novel drugs and discovery of drug targets for disrupting biofilms, leading to the effective improvement of clinical treatment of antibiotics-resistant bacterial pathogens and biofilm-related bacterial chronic infections.

2. Bacterial biofilm formation and regulation

In specificity, bacterial biofilm is a highly complex, well-organized, three-dimension-structural consortium of bacteria that are embedded in a self-produced extracellular matrix, containing polysaccharides, proteins and nucleic acids, and so on [10]. Despite the structural complexity of bacterial biofilms, a classical five-step model was previously proposed to explain its formation: (1) reversible attachment phase, (2) irreversible attachment phase, (3) extracellular polymeric substances (EPS) production, (4) maturation, and (5) dispersal and detachment [11]. Recently, Sauer et al. have revised the conceptual model and proposed a simple three-step biofilm formation model, that is, aggregation, growth, and disaggregation, in order to represent a broader range of biofilm system [12]. Biofilm formation is sophisticatedly regulated and is involved in complex network of regulatory cascades such as quorum sensing (QS) system (communications of bacterial cells within biofilm), regulatory small RNAs (sRNAs), second messengers (cyclic-di-guanosine monophosphate, c-di-GMP), and so forth [13], while elucidation of the regulatory mechanisms of biofilm formation will promote the development of effective strategies to biofilm inhibition and control [14].

3. Bacterial biofilm prevention, inhibition, and eradication

Biofilm infections are persistent and recalcitrant, are tightly associated with the rise of antibiotic resistance, and show heterogeneous features with diverse nature [15]. Due to the harmfulness of bacterial biofilms in human infections, effective strategies for preventing, inhibiting, and eradicating biofilms are urgently needed. As the proverb runs, prevention is better than cure [16]. Therefore, it is always the priority to prevent the formation of bacterial biofilm rather than to inhibit and eradicate it, which requires less effort on biofilm control. Multiple strategies are currently available for biofilm prevention, which most frequently involve treating abiotic surfaces (smoothness, wettability, or hydrophilicity) and coating surfaces (salivary proteins, 2-methacryloyloxyethyl phosphorylcholine, monomeric trimethylsilane, antimicrobial peptide) [17]. These methods are able to greatly reduce microbial attachment to device surface, hence preventing biofilm formation and reducing bacterial infection. As for the inhibition of bacterial biofilms, there are also many effective combating tactics like quorum-sensing blockage, hindering the biosynthesis of N-acyl-homoserine lactones (AHL) signal molecule, biodegradation or alteration of AHL signal molecule, interference with receptor proteins by analog compounds, and so on [18]. Further in-depth studies are needed to elucidate the effects and mechanisms of these biofilm inhibition tactics so that they could be used in the host, proving their applicability to humans in clinical settings. Although prevention and inhibition of bacterial biofilms provide some clinical premise, these methods do not represent a direct treatment for established biofilms while eradication agents and approaches are efficient to remove mature biofilms [19]. Several representative methods include electrochemical methods, antimicrobial compounds, biofilm architecture modulation, and drug delivery methods, all of which aim to eradicate bacterial biofilms when applied alone or synergistically [20]. However, the eradication of mature bacterial biofilms is extraordinarily difficult. More strategies and novel compounds need to be developed for a more effective fight against biofilms.

4. Summary

With the development of prevention, inhibition, and eradication methods of bacterial biofilms, the mortality rate of chronic and fatal bacterial infections is expected to be greatly reduced in future.

Acknowledgements

This study was financially supported by Research Foundation for Advanced Talents of Guandong Provincial People's Hospital [Grant No. KY012023293].

Author contributions

LW and BG conceived the framework of the manuscript, provided platform and resources, and contributed to project administration and funding acquisitions. All the authors contributed to the writing and revision of the manuscript and approved the submission of the final version of the manuscript.

Declaration of interest

The authors declare that the research was conducted in the absence of any commercial or financial relationships that could be construed as a potential conflict of interest.

Inclusion and diversity

We support inclusive, diverse, and equitable conduct of research.

Author details

Liang Wang[1,2]* and Bing Gu[1]*

1 Laboratory Medicine, Guangdong Provincial People's Hospital (Guangdong Academy of Medical Sciences), Southern Medical University, Guangzhou, Guangdong Province, China

2 Center for Precision Health, School of Medical and Health Sciences, Edith Cowan University, Joondalup, WA, Australia

*Address all correspondence to: healthscience@foxmail.com and gubing@gdph.org.cn

References

[1] Becker K, Hu Y, Biller-Andorno N. Infectious diseases – A global challenge. International Journal of Medical Microbiology. 2006;**296**:179-185

[2] Baker RE, Mahmud AS, Miller IF, Rajeev M, Rasambainarivo F, Rice BL, et al. Infectious disease in an era of global change. Nature Reviews Microbiology. 2021;**20**:193-205

[3] Wang H, Abbas KM, Abbasifard M, Abbasi-Kangevari M, Abbastabar H, Abd-Allah F, et al. Global age-sex-specific fertility, mortality, healthy life expectancy (HALE), and population estimates in 204 countries and territories, 1950-2019: A comprehensive demographic analysis for the Global Burden of Disease Study 2019. The Lancet. 2020;**396**:1160-1203

[4] Drexler M. What you Need to Know about Infectious Disease. Washington (DC): National Academies Press (US); 2011

[5] Doron S, Gorbach SL. Bacterial infections: Overview. In: International Encyclopedia of Public Health. Academic Press; 2008. pp. 273-282

[6] Sakai T, Morimoto Y. The history of infectious diseases and medicine. Pathogens. 2022;**11**:1-12

[7] Aminov RI. A brief history of the antibiotic era: Lessons learned and challenges for the future. Frontiers in Microbiology. 2010;**1**:1-7

[8] Uruén C, Chopo-Escuin G, Tommassen J, Mainar-Jaime RC, Arenas J. Biofilms as promoters of bacterial antibiotic resistance and tolerance. Antibiotics. 2020;**10**:1-36

[9] Bjarnsholt T. The role of bacterial biofilms in chronic infections. APMIS. 2013;**121**:1-58

[10] Høiby N, Ciofu O, Johansen HK, Song ZJ, Moser C, Jensen PØ, et al. The clinical impact of bacterial biofilms. International Journal of Oral Science. 2011;**3**:55-65

[11] Muhammad MH, Idris AL, Fan X, Guo Y, Yu Y, Jin X, et al. Beyond risk: Bacterial biofilms and their regulating approaches. Frontiers in Microbiology. 2020;**11**:1-20

[12] Sauer K, Stoodley P, Goeres DM, Hall-Stoodley L, Burmølle M, Stewart PS, et al. The biofilm life cycle: Expanding the conceptual model of biofilm formation. Nature Reviews Microbiology. 2022;**20**:608-620

[13] Schulze A, Mitterer F, Pombo JP, Schild S. Biofilms by bacterial human pathogens: Clinical relevance—Development, composition and regulation—Therapeutical strategies. Microbial Cell. 2021;**8**:28-56

[14] Samrot AV, Abubakar Mohamed A, Faradjeva E, Si Jie L, Hooi Sze C, Arif A, et al. Mechanisms and impact of biofilms and targeting of biofilms using bioactive compounds—A review. Medicina. 2021;**57**:1-28

[15] Da Silva RAG, Afonina I, Kline KA. Eradicating biofilm infections: An update on current and prospective approaches. Current Opinion in Microbiology. 2021;**63**:117-125

[16] Borysiewicz L. Prevention is better than cure. Clinical Medicine. 2009;**9**:572-583

[17] Yin W, Xu S, Wang Y, Zhang Y, Chou S-H, Galperin MY, et al. Ways to control harmful biofilms: Prevention, inhibition, and eradication. Critical Reviews in Microbiology. 2020;**47**:57-78

[18] Srinivasan R, Santhakumari S, Poonguzhali P, Geetha M, Dyavaiah M, Xiangmin L. Bacterial biofilm inhibition: A focused review on recent therapeutic strategies for combating the biofilm mediated infections. Frontiers in Microbiology. 2021;**12**:1-19

[19] Verderosa AD, Totsika M, Fairfull-Smith KE. Bacterial biofilm eradication agents: A current review. Frontiers in Chemistry. 2019;7:1-17

[20] Wolfmeier H, Pletzer D, Mansour SC, Hancock REW. New perspectives in biofilm eradication. ACS Infectious Diseases. 2017;**4**:93-106

Chapter 2

Formation, Regulation, and Eradication of Bacterial Biofilm in Human Infection

Muhammad Usman, Huan Yang, Jun-Jiao Wang, Jia-Wei Tang, Li-Yan Zhang and Liang Wang

Abstract

Microbial biofilms are complicated structures in which planktonic cells change to a sessile form of growth. The development of an extracellular polymeric substance (EPS) matrix, which encloses the bacterial cells and offers additional protection, supports that kind of growth. Biofilms present a significant threat to public health due to their extreme resistance to higher antibiotic concentrations. In addition, biofilms are also resistant to human immune systems. Bacterial biofilms can spread their pathogenicity through a variety of approaches, such as adhering to a solid surface, evading host defenses like phagocytosis, generating a large amount of toxins, resisting anti-microbial agents, transferring genes to generate more virulent strains, and dispersing microbial aggregates that transport the microorganisms to new locations. Consequently, there is an urgent need to replace the widespread procedure of antibiotics with novel developing approaches. Furthermore, biofilm formation has been connected with high rates of disease, health-related infections, and even death, leading to the search for alternative treatment approaches. The review intends to provide information about clinically important bacterial pathogens of the gut, mouth, skin, and lungs and insights into the different perceptions of microbial biofilms, as well as their formation, regulation, and pathogenicity. In addition, for efficient eradication or inhibition of biofilms and associated infections, nanoparticle approaches for addressing persistent bacterial infections have also been discussed.

Keywords: biofilm, bacterial infection, pathogenicity, antibiotic resistance, nanoparticles

1. Introduction

The term "biofilm" refers to a connection of microorganisms when microbial cells adhere to one another on living or inactive surfaces and are enclosed in an extracellular polymeric substance (EPS) matrix [1]. The initial identification of microbial biofilm belongs to a Dutch researcher Antoni van Leeuwenhoek, who used a simple microscope to detect "animalcules" for the first time on the surfaces of teeth [2]. A number of studies have demonstrated that bacterial biofilms are resistant to antibiotics and cannot be hindered by human immune system because the microbes that cause biofilms have a greater capacity to resist or remove antimicrobial agents, which extends the period of recovery during infection [3]. During biofilm-forming stage, certain genes of bacteria are induced, resulting in the activation of stress-related genes and the transformation of bacteria into resistant phenotypes which lead to changes in cell density, pH, osmolarity, or nutrition [4]. It has been reported that majority of bacteria have the ability to develop biofilm on almost every type of surface, which poses a significant threat to the health of people because of the diseases it causes and the resistance it provides to many antibiotics [4, 5]. According to studies, the exopolymer in biofilms inhibits the ability of leucocytes to pass through the biofilm, checking their capacity of leucocytes to degranulate, and stops them from producing reactive oxygen species (ROS), which prevents bacterial phagocytosis [6–8]. Previous investigations have stated that significant amounts of clinically important bacterial pathogens such as *Enterobacter cloacae, Escherichia coli, Klebsiella pneumonia*, *Pseudomonas aeruginosa, Staphylococcus aureus, S. epidermidis,* etc. possess the ability to develop biofilms [4, 9–11]. In addition, biofilms are also known to spread diseases through colonizing surgical instruments, which include central venous catheters, urinary catheters, joint prostheses, pacemakers, etc. [11, 12]. Furthermore, biofilm has been associated with chronic wounds, lung infections in cystic fibrosis patients, and dental caries [13].

In this chapter, we will describe the clinically important bacterial pathogens of the mouth, gut, lungs, and skin. Moreover, the formation and regulation of bacterial biofilm as well as pathogenicity, its mechanisms, and the eradication of biofilm by nanoparticles will also be discussed. This information echoes advancements in microbiome diagnostics and shows how biofilm is formed and regulated. A closer examination of biofilm provides more clarity on the inherent strengths and weaknesses of biofilm. It highlights the need to realize that biofilm is not simply a more significant number of wound pathogens but a sophisticated biological process that requires specific, targeted care. This study also highlights how different types of nanoparticles help in the eradication of bacterial biofilm and shows that nanoparticles have an excellent capacity for the eradication of bacterial biofilm and that different types of nanoparticles act in different ways in order to eradicate biofilm.

2. Body-site infection of clinically-important bacterial pathogens

The clinical importance of different bacterial pathogens is widely recognized, and regular examination is required to provide an accurate diagnosis for specific kinds of infections. Some clinically important bacterial pathogens that cause health complications worldwide and occur in the mouth, skin, lungs, and gut are given below.

2.1 Mouth

The mouth, which serves as a pathway to the digestive system, offers a habitat for a diverse and abundant microbial population, and masses of these organisms and their products develop on the surfaces of the teeth and gums [14]. These growths, commonly referred to as plaque from the mouth and classified as biofilms, contribute to the development of cavities, which leads to tooth damage [15]. The microbial community that grows around teeth is extremely complicated. The microbiome of the mouth poses a threat to maintaining overall and dental health [16]. Therefore, dental caries is a dietary-microbial disease that involves a cariogenic biofilm and continuous exposure to fermentable carbohydrates from dietary sources, such as sucrose, glucose, fructose, maltose, etc. [17]. Around 700 different types of bacteria in the mouth cavity have been detected using ribosomal identification techniques, among which *Streptococcus mutans* is one of the most common caries-causing bacterium [18]. It canmetabolize many types of carbohydrates, generating high-level acidity and dextran that facilitates the production of dental plaque. Many different bacterial species belonging to the genera *Streptococcus* and *Actinomyces* can be found in the plaque biofilms [19, 20]. For example, the two Gram-positive and anaerobic bacteria, *Streptococcus anginosus* and *Actinomyces naeslundii*, are commonly found in biofilms, while under healthy circumstances, Gram-negative bacteria such as *Aggregatibacter Actinomycetemcomitans*, *Campylobacter* spp., *Porphyromonas spp.*, *Prevotella intermedia,* and *Treponema denticola* can also exist [21]. The investigations have shown that these Gram-negative bacteria might infect other parts of the body when hygiene in the mouth fails to be observed. For example, *staphylococci* (*staph*) and *streptococci* (strep are involved in endocarditis) [22]. In addition to this, other issues related to these biofilms include actinomycosis, dental root infections, and foul breath [23].

2.2 Skin

The human skin microbiome performs a significant role in both health and disease. The initial defense line of human body from pathogens is the skin, which protects and shields the body and provides a hostile environment for majority of bacteria [24]. Microbial biofilms are an extensively investigated mode of surface-associated growth that exhibits community-like characteristics. Furthermore, biofilms play an important role in numerous skin diseases. The usual microbiota of skin comprises a considerable number of Gram-positive bacteria, such as *Staphylococci* and *Micrococci*. Gram-positive bacteria are comparatively resistant to harsh conditions such as dryness and extreme osmosis pressures noticed in high salt or sugar mixtures [25]. The common causes of bacterial infections in skin are *Staphylococcus* and *Streptococcus* [26]. The bacteria *Streptococcus pyogenes* is responsible for a contagious bacterial skin infection that forms pustules and yellow, crusty sores. In certain instances, both *S. aureus* and *S. pyogenes* are present. Usually, the bacteria that cause infection penetrate through a small skin opening. Additionally, the infection has the potential to spread to neighboring body parts. However, the primary skin pathogens are coryneform bacteria like hemolytic *Streptococci* and *S. aureus* [27]. Normally, such bacteria penetrate the body from a wound in the skin, including bites from insects, etc. [28]. *In vitro,* single-species biofilms of skin microbiota, such as *S. aureus*, *S. epidermidis*, and *Propionibacterium acnes*, have been investigated [29]. Furthermore, important inter-species interactions with skin prokaryotes have been found, e.g., *S. epidermidis* inhibiting both *P. acnes* growth and *S. aureus* biofilms [29, 30].

2.3 Lungs

Lung infection is one of the most prominent health issues. Various systemically or respiratory problems start with lung infections. The region of the upper respiratory tract is where airborne pathogens initially come into interaction with the body's mucous membranes [31]. Pathogens from surroundings and dust particles are continuously exposed to the pulmonary system and airways. Once there are issues with any component of this system, lung disease can occur [32]. The most common causes of bacterial lung infections in normal hosts include *Streptococcus pneumoniae, Haemophilus species, Staphylococcus aureus, and Mycobacterium tuberculosis*. The bacterial infections of the lungs were responsible for one-fifth of all fatalities in Europe and North America between the seventeenth and nineteenth centuries. It can remain dormant for years before establishing a chronic cavitating lung infection with highly infectious sputum. Following a significant reduction in prevalence, mainly due to advancements in public health, *M. tuberculosis* infections are currently reducing in rate, while multidrug-resistant strains are spreading across various areas [33, 34]. Other common microorganisms responsible for pneumonia include *Staphylococcus aureus, Group A Streptococcus, Klebsiella pneumoniae, Haemophilus influenzae,* and *Moraxella catarrhalis* [35]. Additionally, patients with smoking-related lung disease frequently have *Haemophilus influenzae* infection, which can result in bronchial inflammation and patchy infiltration into the surrounding lung. *H. influenzae* is under-detected by the usual clinical culture approach [36].

2.4 Gut

Bacteria can enter the intestinal mucosa and replicate there, as well as spread to other organs in the body [37]. The majority of intoxications, including those carried on by *Staphylococcus aureus*, are identified by the symptoms appearing extremely quickly (often within a few hours) [38]. The gut microbiota is a convoluted ecology with approximately 300 to 500 different bacterial species [39]. In contrast to the lower gut, the stomach and upper intestine have less abundant microbiota [40]. Bacteria can be found in the mucosa and in the lumen, although they often are unable to penetrate the gut wall. The usual intestinal microbiota contains small populations of bacteria which can lead to disease when allowed to overgrow. For instance, an over-population of *Clostridium difficile* can result in serious intestinal inflammation and diarrhea. Antibiotic administration starts the procedure by inhibiting the natural microbiota [41]. The main common pathogens of the gut include *Vibrio cholerae,* enteropathogenic strains of *E. coli, Eubacterium, Bacillus cereus, Bacteroides vulgatus, Bifidobacterium, Clostridium difficile, Fusobacterium, Peptostreptococcus, Pseudomonadota, Prevotella, Salmonella enterica, Salmonella gastroenteritis, Salmonella typhimurium,* and *Shigella spp* [42–48]. *Bacillus subtilis* is a gut commensal and non-pathogenic [49, 50]. *Lactobacillus johnsonii* and *Clostridium perfringens* are both commensal [51, 52]. However, *Clostridium perfringens* is also an opportunistic pathogen that can lead to lethal diseases as a result of overgrowth causing gas gangrene, food poisoning, non-foodborne diarrhea, and enterocolitis [53, 54]. *Bacteroides fragilis* is part of the normal microbiota of the human colon and is commensal, but can cause infection if displaced into the bloodstream or surrounding tissue following surgery, disease, wounds, or trauma [55, 56].

3. Biofilm formation and regulation

Wound contamination happens within minutes when planktonic (free-swimming) microorganisms travel into the wound, anchor to the wound bed, and become attached (sessile) [57]. Bacteria produce sticky sugar strands or polymers known as extracellular polymeric substances (EPS) when attached to the wound bed. These polymers form bonds with the help of metallic ions obtained from the host and wound environment, forming a three-dimensional protective structure that grows into a complex community developed to protect the encased bacteria compared to assault through the body's immune system or external attack.

3.1 Steps of biofilm formation

The development of a biofilm actually involves a combination of physical, chemical, and biological processes that proceed over a period of time. Detailed procedures of bacterial biofilm formation are illustrated in **Figure 1**.

3.1.1 Conditioning

The formation of a conditioning layer is the initial stage in the development of a biofilm. The components of the bulk fluid settle onto the surface at this stage, developing a substratum. In common, rough surfaces and hydrophobic materials have a preference compared to hydrophilic and smooth surfaces [59]. The microorganism adheres to such surfaces, which are consequently changed to improve a surface charge that assists in the attraction and adhesion of bacteria with opposing charges [60]. The bacteria can adhere to the surface more strongly owing to the existence of pili, fimbriae, and glycocalyx on the surface [61]. Although an initial adhesion can be reversed, if attraction dominates over repulsion, it will become irreversible.

3.1.2 Attachment and growth

When the adhesion is effective, the bacteria start to grow by taking advantage of the available nutrients. Following this stage, biological events dominate bacterial adhesion to the surface. This is the outcome of the expression of a number of genes that are in the position of producing surface proteins such as porins [62].

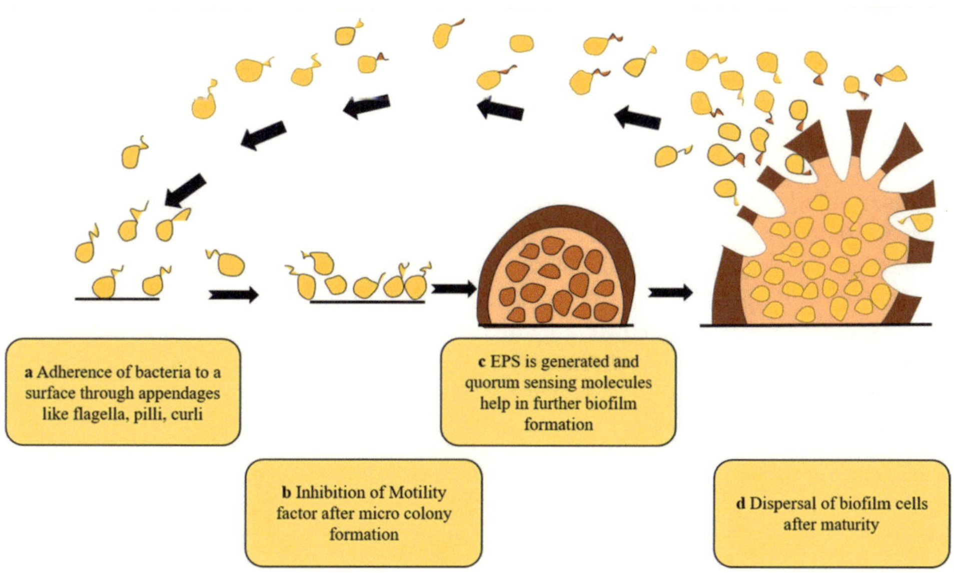

Figure 1.
Schematic illustration of the key steps in biofilm formation: (a) planktonic bacteria attaching to the surface, (b) motility factor inhibition, (c) extracellular polymeric substances (EPS) generation, and (d) biofilm maturation and dispersal. The figure is adapted and modified from a previous study with copyright permissions [58].

The polysaccharides used to create the EPS layer are transported with the help of porins. Since the biofilm matures, microbial cells start to connect to each other via the release of autoinducer signals (AIs) [63]. This communication is critical due to an established biofilm can comprise up to 100 billion bacterial cells per milliliter. The cells are divided into identical different groups, each of which is accountable for a specific task [64]. Another frequently observed phenomenon in a growing biofilm is the formation of high and wrinkled structures. However, this causes lateral pressure on the cells by pulling it towards one another. The dead cells in the biofilm concentrate in the areas that promote vertical bulging, which helps in releasing this pressure [65].

3.1.3 Metabolism

The metabolic process of the biofilm modifications with changes in the environment of the biofilm throughout the primary phase of growth of biofilm, when the metabolic activity is strong and subsequently declines as growth progresses [66]. The complex diffusion channels are employed as the cell population grows to transport nutrients, oxygen, and further components required for cell growth. These channels are used to transport the metabolic wastes and debris. In fact, shear stress has a significant impact on the expression of genes involved in glycolysis [67]. The bacteria that form biofilms have a propensity to ingest foreign DNA, which could eventually lead to the expression of exogenous proteins [68]. Furthermore, it has been demonstrated that several genes involved in the biosynthesis of fatty acids were downregulated as the biofilm formed [69]. These findings show that biofilm-forming cells have a very different metabolism from planktonic cells.

3.1.4 Dispersion

The final step is dispersion, which involves the destruction of the biofilm and the sessile cells, allowing them to resume their motile forms. Finally, biofilm makes use of its disruptive forces to spread throughout, and the bacteria colonize new regions and develop [70].

3.2 Biofilm regulation

This protective biofilm structure is comprised of proteins and smaller molecules which are strung together to form larger, more robust polymer units of sugars (polysaccharides), macromolecules like DNA, and lipids [71]. This EPS helps the bacteria contained within the structure survive by supplying nutrients, removing waste products, and preventing harmful antimicrobial molecules, antibodies, and host inflammatory cells from getting or interacting with the bacteria. A developed biofilm additionally helps in the ongoing maturation and eventual spreading recolonization of the encased bacteria, prevents molecules too large to pass through the structure, and provides a diffusion barrier to small molecules like antibiotics (**Figure 2**) [4, 73–76].

Defensive EPS structure of bacterial biofilm protects bacterial hybridization, tolerance, and gene expression by subverting the natural infection inflammatory response in order to get rid of the body of bacteria and support the survival of bacteria [75]. It is noteworthy that this protective structure can repel treatments and promote continued biofilm growth, even if the biofilm is chemically or mechanically fractured into microcolonies, rendering the bacteria within the structure virtually invincible unless the structure is solubilized and removed. The protective EPS structure of the

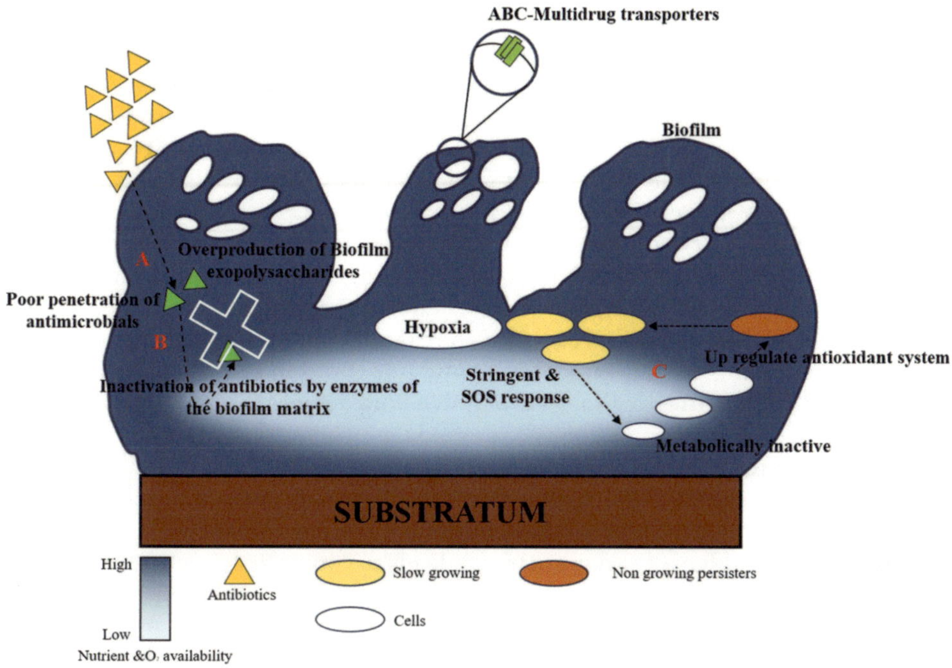

Figure 2.
The general mechanism of biofilm resistance to antimicrobials. (A) Biofilm matrix provides a diffusion barrier to small molecules like antibiotics. (B) Inactivation of antibiotics by enzymes of the biofilm matrix. (C) Persister cells in the deeper layer of biofilm inducing adaptive SOS reaction and hence developing further resistance. The figure is adapted and modified from a previous study with copyright permissions [72].

biofilms protects bacterial hybridization, tolerance, and gene expression patterns and promotes bacterial survival by preventing the body's own inflammatory response that is designed to get rid of bacteria [18]. This procedure enables the biofilm to rapidly mature and develop impenetrable to conventional treatments as well as unculturable using conventional culture methods. The biofilm may function passively as a reservoir for pathogenic bacteria which are typically polymicrobial in nature, or it can take a more active role by encouraging an expanding area of inflammation and pathogenic tissue damage that favors the progression into overt infection as the biofilm develops into a more mature insoluble biomass, encouraging protected bacterial growth, mutation, and proliferation through sophisticated cell-to-cell and cell-to-surface interactions between the host and the biofilm [77–81]. Over time, a portion of the biofilm's bacteria disperses as fresh-roving bacteria and micro-bacterial aggregates that release and spread, acting as the foundation for new biofilm colonies [82].

4. Bacterial biofilms: Pathogenicity and properties of the bacterial biofilms

4.1 Pathogenicity

It is well-established that biofilms contribute to the virulence of pathogens. According to statistics from the Centres for Disease Control and Prevention (CDC) and the National Institutes of Health (NIH), the prevalence of disease caused by biofilms is believed to be between 65% and 80%, particularly in developed nations

[83]. Several food-borne pathogens including *E. coli, Salmonella, Yersinia enterocolitica, Listeria,* and *Campylobacter* create biofilms on the surface of food or storage equipment. Furthermore, potentially pathogenic bacteria such as *Staphylococcus aureus, Enterococcus faecalis, Streptococcus, E. coli, Klebsiella,* and *Pseudomonas* thrive on catheters, prosthetic joints, mechanical heart valves, and so on. As a result of their periodic escape from the said focus these organisms may cause persistent diseases [83, 84]. The localized depletion of nutrition in a biofilm has been proposed as an inducer of cell release or detachment from the biofilm in *Pseudomonas aeruginosa* [85]. However, microbially produced gas bubbles, cross-linking cations, growth status, contact surface material, shear stress, quorum sensing, and lytic bacteriophage activation have all been identified as major contributors to biofilm detachment. They can be life-threatening causing endocarditis and infections in people with cystic fibrosis, in addition to infecting long-term indwelling devices like heart valves and joint prostheses [86].

Numerous bacterial toxicities in the human body such as the development of dental plaques, infections of the middle ear in children, urinary tract infections, gingivitis, and contact lens infections are caused by biofilms. The biofilm formation takes place on contact lenses and ultimately it leads to contamination [9, 87].

4.2 Properties of the bacterial biofilms

The bacterial biofilms cause pathogenicity through a variety of unique properties.

4.2.1 Variation in phase

Small colony variations (SCVs) are a colony phenotype highlighted by small size, slow development, and virulence gene downregulation and have been identified as a pathogenic mechanism for various bacterial species such as *S. epidermidis* and are often linked with chronic infections [88]. Biofilms have the unique potential to generate bacterial subpopulations that are shifted to a dormant state and are known as small colony variations (SCVs). They also have low catalase activity which interferes with oxidative metabolism. SCVs generate noticeable morphological changes in biofilms, increasing adhesion, auto-aggregation, hydrophobicity and pathogenicity. These variations contribute to biofilm survival in extreme environmental conditions [58]. SCVs appear to be less sensitive to antibiotics and the immune system possibly due to their ability to survive intracellularly and induce a more anti-inflammatory setting due to higher Inter Leuken-10 (IL-10) release [89].

4.2.2 Efflux pumps

Efflux pumps are present in the periplasmic space within the bacterial membranes and have a negative influence on antibiotic accumulation and their presence is critical in pathogenesis in biofilm [90]. The mutations in regulation proteins or promoters result in the production of these efflux pumps which causes pathogenicity. These efflux pumps are energy-dependent and on the basis of mechanism by which they derive energy are generally categorized into two groups. The primary efflux pumps get their energy from constant hydrolysis of ATP whereas the secondary efflux pumps get their energy from chemical gradients created by protons or ions like sodium ions [91]. The increased expression of these efflux pumps in biofilm has also been linked to pathogenesis in *P. aeruginosa* biofilms by causing antibiotic resistance [92].

4.2.3 Alterations in membrane protein expression

The outer membrane channel proteins existing in bacterial membranes play an important part in transferring hydrophilic particles from the outer atmosphere to the periplasmic space in biofilm of bacteria [93]. The presence of outer membrane proteins (Omps) allows for macromolecular interaction between the cell and the environment. These proteins are established in the outer membrane of bacteria in biofilm. A larger channel outer membrane protein such as OmpF can be used in place of OmpC because it has a smaller diameter. This change limits the entry of bigger compounds with high hydrophobicity such as carbenicillin. In contrast, small hydrophilic molecules such as imipenem can pass through the OmpC channels. In biofilm, differential expression of outer membrane protein-coding genes occurs which contributes to antibiotic resistance and pathogenicity [94].

5. Bacterial biofilm and its eradication

Many types of nanoparticles with therapeutic effects against bacterial biofilm infections can be categorized according to their chemical composition or their healthcare purposes. The nano-formulations have a high attraction to the bacterial cells and the capacity to penetrate biological barriers like biofilm because of their small size, large surface area, and highly sensitive nature [95]. The sizes of the NPs are sufficiently small to penetrate into biofilms and microbial cell walls, while the large surface area of the NPs enables the effective loading of drugs [96]. Although the exact strategy of NPs reducing biofilm formation is not completely elucidated yet, multiple studies have described different processes by which NPs impact bacterial cells and biofilms (**Figure 3**). There are various common types of nanoparticles used for biofilm eradication, which are listed and discussed below.

5.1 Metal-based nanocomposites

Metal nanoparticles (MNPs) are often employed in antibacterial and antibiofilm studies due to their inherent nature, structure, and large surface-to-ratio, enabling control in fabrication, approach, and modification of their physical and chemical characteristics [98]. MNPs demonstrate significantly greater antibacterial activity compared to their micro-sized counterparts, although NPs, like common antimicrobials, lack the ability to recognize sensitive and resistant microorganisms [96]. However, their non-specificity is also one of their drawbacks because they can also attack commensal bacteria [99]. Metal oxide nanoparticles (MONPs), silver nanoparticles (AgNPs), gold nanoparticles (AuNPs), and various other metal-based nanocomposites (NCs) have shown their efficiency in preventing biofilm formation through a distinct inhibitory mechanism [96, 99].

5.1.1 Metal oxide NPs

Metal oxide nanoparticles (MONPs) that demonstrated antimicrobial activity consist of iron oxide (Fe_3O_4), zinc oxide (ZnO), titanium oxide (TiO_2), silicon oxide (SiO_2), selenium oxide (SeO_2), and aluminum oxide (Al_2O_3). The majority of the NPs impacts on microbial cells involve cellular membrane breakdown resulting from NP-cell surface interaction and consequent leak of cell substance [100]. Metal oxide

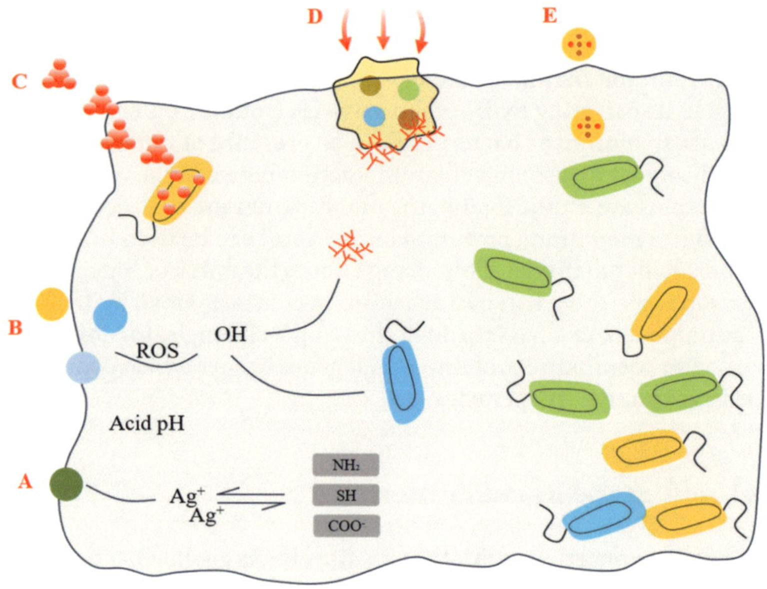

Figure 3.
The primary mechanisms of nanoparticles (NPs) in biofilm eradication. (A) Interaction with functional components of biofilm via the released ions. (B) Production of reactive oxygen species (ROS) that cause bacterial destruction and EPS breakdown. (C) Antimicrobial-loaded polymeric nanoparticles penetrate into the biofilm and deliver drugs to the bacterial cell. (D) the photothermal effect, which occurs in the presence of near-infrared (NIR) light irradiation, causing an increase in local heat, which acts efficiently alongside EPS and bacterial cells. (E) Liposomes encapsulate the antimicrobial and fuse with cell membranes, allowing the antibiotic to be released directly inside the bacterial cell. The figure is adapted and modified from a previous study with copyright permissions [97].

NPs (MONPs) also mediate mechanisms for DNA and RNA destruction, the production of ROS, and the discharge of poisonous substantial metal ions [98]. The primary antibacterial activity of these NPs is related to oxidative stress, which is caused by the formation of ROS on the outer layers of metal oxides and subsequent breakdown of cell membranes, structure of cells, and molecules [98, 101]. The effects of oxidative stress can harm the proteins that contribute to attachment and biofilm development. Furthermore, it suppresses the development of genes that are important for bacterial cell attachment on surfaces and biofilm development [102].

5.1.2 Silver nanoparticles (AgNPs)

AgNPs are commonly used as antimicrobial agents and have greater antibacterial activity than some antibiotics as well as employed in clinically developed devices, tubes, and dressings for wounds [103]. In addition to exhibiting antimicrobial properties, AgNPs possess a large surface-to-mass ratio, which makes them an attractive choice for use as single layers on the surfaces of biomolecules [104]. The antibacterial effect of AgNPs is considered to be caused by the NPs' breakdown and the release of Ag^+ ions, which attach to the cell membrane and depolarize the cell wall while changing the permeability and negative charge of the membrane. Additionally, when Ag^+ ions penetrate the target bacterium, they cause the oxidation and breakdown of cellular components, the reduced activity of respiratory chain enzymes, and the formation of ROS, which hinders the recombination of DNA and the fabrication of ATP [98, 105].

5.1.3 Gold nanoparticles (AuNPs)

AuNPs are more efficient against biofilm compared to AgNPs because they have a lower hydrophobicity index, which reduces the growth of biofilm [106, 107]. The antimicrobial process of AuNPs is believed to involve affecting the membranes of bacterial cells and inhibiting ATPase production, which leads to metabolic degradation, as well as hindering the ribosome component attaching to tRNA, attacking nicotinamide, and impacting the bacterial respiratory chain [105]. In addition to having antimicrobial qualities, AuNPs also exhibit photothermal characteristics when exposed to near-infrared (NIR) light. This is because accumulated AuNPs absorb light in a red-shifted manner, which causes a dramatic increase in localized heat. Consequently, this represents yet another potent means of eradicating bacteria from the infectious biofilm without damaging the tissues that surround it because their cells require a greater amount of heat due to their larger size than bacterial cells [108, 109].

5.1.4 Metal-polymer nanocomposites (MNCs)

The fabrication of MNPs as polymer nanocomposites increases their stability and efficiency. According to Nagvenkar et al. [110], adding ZnO NPs to polyvinyl alcohol (PVA) polymer enhanced the stability and efficiency of the ZnO-PVA nanofluid against *S. aureus and E. coli* [110]. The toxic effects of ZnO can be decreased through its incorporation into different materials. Banerjee et al. [111] showed that doping pancreatin (PK) on ZnONPs (ZnONPs-PK) reduced the toxic effects of ZnO and increased its antibacterial and anti-biofilm efficiency while decreasing its virulence towards MRSA [111]. Depan and Misra [112] found that incorporating titania NPs into silicone decreased *S. aureus* life and adhesive abilities by 93% when compared to stand-alone silicone [112]. Silicone-TiO_2 NPs were also more effective at breaking down the biofilm; after 6 hours of incubation, the biofilm completely disintegrated. Wang et al. [113] observed that AuNPs incorporated with graphitic carbon nitride (g-C3N4) could improve H_2O_2 efficiency by causing peroxidase-like activity that effectively breaks down H_2O_2 to OH radicals, leading to demonstrated biofilm destruction and prohibited biofilm development in vitro, reducing the growth of *E. coli and S. aureus*, and potentially accelerating the healing process [113]. Recently, capping MNPs with polysaccharides obtained from other microbes like yeast and algae has been shown to be effective [114].

5.2 Polymer-based nanoparticles (PNPs)

Polymer nanoparticle-based antimicrobial delivery systems are whatever they are commonly referred to as in regards to their functionality. Despite the fact that the chemical composition could be organic, inorganic, or even a mixture of both, their enhanced antibacterial transport is due to their improved stability, capacity for modification, formation at the site of infection, and monitored release ability, along with boosted cytocompatibility and biodegradable properties [98, 115, 116]. PNPs have a distinct advantage over MNPs because medications can be maintained within their cavity, allowing the drug to be delivered to the target region, whether confined or entrapped [99, 117]. Based on this, PNPs are available in two shapes, nanospheres and nanocapsules, with sizes ranging from 100 to 500 nm. The nanosphere is a polymeric matrix that contains the drug that has been adsorbed in it. The drug can be entrapped in small cavities or adsorbed onto the polymer wall of nanocapsules, which have an oily core and a polymeric shell around them [117].

5.3 Natural and synthetic polymer-based nanoparticles (PNPs)

PNPs can be either synthetic or natural, like chitosan, polycaprolactone, polylactic acid, and polylactic-co-glycolic acid (PLGA). Chitosan, a cationic heteropolysaccharide, is frequently used as a nanocarrier due to its biocompatibility, immunostimulating properties, non-toxicity, biodegradability, adhesive properties, and relatively low cost of production [115, 118]. Chitosan has a high ability to inhibit biofilm growth because of its polycationic nature, which results in electrostatic interaction with the biofilm components and damages the biofilm matrix [119]. Additionally, electrostatic interaction between positively charged chitosan and negatively charged bacterial cell surfaces leads to the destruction of bacterial cell membranes and the leakage of their constituent parts, as well as the inhibition of mRNA transcription and protein synthesis through DNA binding [100].

5.4 pH-responsive polymer-based nanoparticles

The ability to be functionalized in accordance with the conditions of the microenvironment determines the polymeric NPs. As previously discussed, the rapid pH-responsive transmission of the NPs negative to positive charge increases their capacity to accumulate and penetrate biofilms inside acidic microenvironments, which decreases drug efficacy. In addition to effectively binding to the bacterial cell surfaces and enhancing photoinactivation efficiency against Gram-negative bacteria, the pH-responsive polymeric NPs carriers have a high potential to interact with the acidic biofilm microenvironment and respond to pH variation [120, 121]. The acidic pH-responsive NPs systems have been developed as bilayers with a cationic outer shell for binding with EPS components and a hydrophobic inner shell for releasing the encapsulated drug and enhancing antimicrobial and antibiofilm activity [120, 122]. For example, Horev et al. [120] conducted an in vitro and in vivo study that demonstrated farnesol-loaded pH-activated polymer NPs had a 4-fold greater ability to inhibit *S. mutans* biofilms than free farnesol; additionally, the drug was concentrated at the biofilm-EPS matrix interface, which greatly improved farnesol retention and bioavailability [120].

6. Conclusion

Human pathogenic biofilms are associated with chronic and recurrent diseases that can be very severe and even fatal. Biofilm formation and regulation are multistep complex procedures that involve the transition of bacteria from free-swimming planktonic form to biofilm-making sessile form. Pathogenicity is the capacity of a pathogen to cause disease through a variety of mechanisms. Pathogenic biofilm may cause different host reactions in a human host and use a variety of mechanisms to evade the host defense systems. Furthermore, toxins such as invasins, lipopolysaccharides (LPS), and cell wall components of biofilm can damage host cells and cause septic shock. Moreover, adhesins help in adhering the pathogen to the surfaces of the host. During infection, bacteria are able to attach to a host surface and continue to penetrate host tissues. Pathogens can "burrow" more deeply into a tissue by generating and releasing proteases and glycanases that degrade host extracellular matrix proteins and polysaccharides. Another possibility for a pathogen in biofilm is to enter the host tissue cells and have access to the intracellular environment. In recent years, nanotechnology has developed into an exciting technique for eradicating bacterial biofilm-related infections. The ability of nanoparticles (NPs) to deliver drugs to the target site in the ideal dosage range, protect them from deactivation, and increase their therapeutic efficiency with fewer side effects makes them a promising therapeutic approach. Aside from that, the small size, large surface area, and highly reactive nature of nanoparticles enable them to penetrate biological barriers like biofilm and have a high eradicative selectivity for bacterial infections. Taken together, this study systematically reviews the formation, regulation, and eradication of bacterial biofilm in human infection, which not only facilitates our understanding of bacterial biofilms but also strengthens the importance of developing novel methods and technologies to inhibit and eradicate biofilm-related human infections, which will greatly reduce the mortality rate of chronic and fatal bacterial infections.

Acknowledgements

We thank the anonymous reviewers for their thoughtful comments that greatly improve the quality of the manuscript.

Author contributions

LW and MU conceived the framework. LW and LYZ provided platform and resources. LW contributed to project administration and student supervision. MU, HY, JJW, JWT contributed to literature review. JWT contributed to the schematic illustration. All the authors wrote and revised the manuscript. All the authors approved the submitted version of the manuscript.

Funding statement

This research was funded by Guangdong Basic and Applied Basic Research Foundation (Grant No. 2022A1515220023) and Research Foundation for Advanced Talents of Guandong Provincial People's Hospital (Grant No. KY012023293).

Conflict of interest

The authors declare that they have no known competing financial interests or personal relationships that could have appeared to influence the work reported in this paper.

Author details

Muhammad Usman[1†], Huan Yang[2†], Jun-Jiao Wang[1], Jia-Wei Tang[3], Li-Yan Zhang[3,4*] and Liang Wang[3,5*]

1 School of Medical Informatics and Engineering, Xuzhou Medical University, Xuzhou, Jiangsu Province, China

2 Medical Technology School, Xuzhou Medical University, Xuzhou, Jiangsu Province, China

3 Laboratory Medicine, Guangdong Provincial People's Hospital (Guangdong Academy of Medical Sciences), Southern Medical University, Guangzhou, Guangdong Province, China

4 Laboratory Medicine, Ganzhou Municipal Hospital, Guangdong Provincial People's Hospital Ganzhou Hospital, Ganzhou, Guangdong Province, China

5 Centre for Precision Health, School of Medical and Health Sciences, Edith Cowan University, Perth, Western Australia, Australia

*Address all correspondence to: zhangliyan@gdph.org.cn and healthscience@foxmail.com

† These authors contributed to the study equally.

References

[1] Hall-Stoodley L, Costerton JW, Stoodley P. Bacterial biofilms: From the natural environment to infectious diseases. Nature Reviews Microbiology. 2004;**2**(2):95-108

[2] Costerton JW, Stewart PS, Greenberg EP. Bacterial biofilms: A common cause of persistent infections. Science. 1999;**284**(5418):1318-1322

[3] Das PK, Samal S. Microbial biofilms: Pathogenicity and treatment strategies. PharmaTutor. 2018;**6**(1):16-22

[4] Fux CA, Costerton JW, Stewart PS, Stoodley P. Survival strategies of infectious biofilms. Trends in Microbiology. 2005;**13**(1):34-40

[5] Khan S, Singh P, Ansari M, Asthana A. Isolation of Shigella species and their resistance patterns to a panel of fifteen antibiotics in mid and far western region of Nepal. Asian Pacific Journal of Tropical Disease. 2014;**4**(1):30-34

[6] Thurlow LR, Hanke ML, Fritz T, Angle A, Aldrich A, Williams SH, et al. Staphylococcus aureus biofilms prevent macrophage phagocytosis and attenuate inflammation in vivo. The Journal of Immunology. 2011;**186**(11):6585-6596

[7] Bayer AS, Speert D, Park S, Tu J, Witt M, Nast C, et al. Functional role of mucoid exopolysaccharide (alginate) in antibiotic-induced and polymorphonuclear leukocyte-mediated killing of Pseudomonas aeruginosa. Infection and Immunity. 1991;**59**(1):302-308

[8] Malic S, Hill KE, Playle R, Thomas DW, Williams DW. In vitro interaction of chronic wound bacteria in biofilms. Journal of Wound Care. 2011;**20**(12):569-577

[9] Parsek MR, Singh PK. Bacterial biofilms: An emerging link to disease pathogenesis. Annual Reviews in Microbiology. 2003;**57**(1):677-701

[10] Miller MB, Bassler BL. Quorum sensing in bacteria. Annual Reviews in Microbiology. 2001;**55**(1):165-199

[11] Ma L, Conover M, Lu H, Parsek MR, Bayles K, Wozniak DJ. Assembly and development of the Pseudomonas aeruginosa biofilm matrix. PLoS Pathogens. 2009;**5**(3):1000354

[12] Wolcott R, Rhoads D, Bennett M, Wolcott B, Gogokhia L, Costerton J, et al. Chronic wounds and the medical biofilm paradigm. Journal of Wound Care. 2010;**19**(2):45-53

[13] Wolcott RD, Rumbaugh KP, James G, Schultz G, Phillips P, Yang Q, et al. Biofilm maturity studies indicate sharp debridement opens a time-dependent therapeutic window. Journal of Wound Care. 2010;**19**(8):320-328

[14] Al Jassim RA, Andrews FM. The bacterial community of the horse gastrointestinal tract and its relation to fermentative acidosis, laminitis, colic, and stomach ulcers, veterinary clinics: Equine. Practice. 2009;**25**(2):199-215

[15] Rosan B, Lamont RJ. Dental plaque formation. Microbes and Infection. 2000;**2**(13):1599-1607

[16] Deo PN, Deshmukh R. Oral microbiome: Unveiling the fundamentals. Journal of Oral and Maxillofacial Pathology: JOMFP. 2019;**23**(1):122

[17] Pitts N, Zero D, Marsh P, Ekstrand K, Weintraub J, Ramos-Gomez F, et al. Dental caries. Nature Reviews. Disease Primers. 2017;**3**:17030

[18] Parahitiyawa N, Scully C, Leung W, Yam W, Jin L, Samaranayake L. Exploring the oral bacterial flora: Current status and future directions. Oral Diseases. 2010;**16**(2):136-145

[19] Kalpana B, Prabhu P, Bhat AH, Senthilkumar A, Arun RP, Asokan S, et al. Bacterial diversity and functional analysis of severe early childhood caries and recurrence in India. Scientific Reports. 2020;**10**(1):21248

[20] Wright PP, Ramachandra SS. Quorum sensing and quorum quenching with a focus on cariogenic and periodontopathic Oral biofilms. Microorganisms. 2022;**10**(9):1783

[21] Rubin M. An in Vitro Model System for Evaluating Substantive Antimicrobial Agents in Simulated Dental Plaque Biofilms. New Jersey: Rutgers The State University of New Jersey-New Brunswick and University of Medicine and Dentistry of New Jersey; 2005:3176220

[22] Vilcant V, Hai O. Bacterial endocarditis. StatPearls. 2017:29262218

[23] Aguado JM, Silva J, Fernández-Ruiz M, Cordero E, Fortún J, Gudiol C, et al. Management of multidrug resistant gram-negative bacilli infections in solid organ transplant recipients: SET/GESITRA-SEIMC/REIPI recommendations. Transplantation Reviews. 2018;**32**(1):36-57

[24] Yoshio H, Lagercrantz H, Gudmundsson GH, Agerberth B. First line of defense in early human life. Seminars in Perinatology. 2004;**28**:304-311

[25] Oprica C. Bacteriology of the skin. Life-Threatening Dermatoses and Emergencies in Dermatology. 2009;**303**:29-36

[26] Becker K, Heilmann C, Peters G. Coagulase-negative staphylococci. Clinical Microbiology Reviews. 2014;**27**(4):870-926

[27] Tognetti L, Martinelli C, Berti S, Hercogova J, Lotti T, Leoncini F, et al. Bacterial skin and soft tissue infections: Review of the epidemiology, microbiology, aetiopathogenesis and treatment: A collaboration between dermatologists and infectivologists. Journal of the European Academy of Dermatology and Venereology. 2012;**26**(8):931-941

[28] Tayade P, Kumawat J, Sunita N, Abhay Joshi JR. Evaluation of antibacterial activity of 3-(1-methoxy napthalen-2-yl)-5-phenylisoxazole. Journal of Medical and Pharmaceutical Innovations. 2016;**3**:2347-8136

[29] Otto M. Staphylococcal biofilms. Bacterial Biofilms. 2008;**322**:207-228

[30] Iwase T, Uehara Y, Shinji H, Tajima A, Seo H, Takada K, et al. Staphylococcus epidermidis Esp inhibits Staphylococcus aureus biofilm formation and nasal colonization. Nature. 2010;**465**(7296):346-349

[31] Wang CC, Prather KA, Sznitman J, Jimenez JL, Lakdawala SS, Tufekci Z, et al. Airborne transmission of respiratory viruses. Science. 2021;**373**(6558):eabd9149

[32] Mao N, An C, Guo L, Wang M, Guo L, Guo S, et al. Transmission risk of infectious droplets in physical spreading process at different times: A review. Building and Environment. 2020;**185**:107307

[33] Liu H, Liu B, Zheng F, Chen X, Ye L, He Y. Distribution of pathogenic bacteria in lower respiratory tract infection in lung cancer patients after chemotherapy and analysis of integron resistance genes in respiratory tract isolates of uninfected

patients. Journal of Thoracic Disease. 2020;**12**(8):4216

[34] Wilson LG. Commentary: Medicine, population, and tuberculosis. International Journal of Epidemiology. 2005;**34**(3):521-524

[35] Pfaller MA, Farrell DJ, Sader HS, Jones RN. AWARE Ceftaroline surveillance program (2008-2010): Trends in resistance patterns among Streptococcus pneumoniae, Haemophilus influenzae, and *Moraxella catarrhalis* in the United States. Clinical Infectious Diseases. 2012;**55**(3):187-193

[36] Kennedy W, Chang S, Purdy K, Le T, Kilgore P, Kim J, et al. Incidence of bacterial meningitis in Asia using enhanced CSF testing: Polymerase chain reaction, latex agglutination and culture. Epidemiology & Infection. 2007;**135**(7):1217-1226

[37] Marshall JC, Christou NV, Horn R, Meakins JL. The microbiology of multiple organ failure: The proximal gastrointestinal tract as an occult reservoir of pathogens. Archives of Surgery. 1988;**123**(3):309-315

[38] Hennekinne J-A, De Buyser M-L, Dragacci S. Staphylococcus aureus and its food poisoning toxins: Characterization and outbreak investigation. FEMS Microbiology Reviews. 2012;**36**(4):815-836

[39] Guarner F, Malagelada J-R. Gut flora in health and disease. The Lancet. 2003;**361**(9356):512-519

[40] Gorbach S. Microbiology of the gastrointestinal tract. Medical Microbiology, University of Texas Medical Branch at Galveston. Retrieved July, 15 2017 1996.

[41] Pickard JM, Zeng MY, Caruso R, Núñez G. Gut microbiota: Role in pathogen colonization, immune responses, and inflammatory disease. Immunological Reviews. 2017;**279**(1):70-89

[42] Burrell M, Hanfrey CC, Murray EJ, Stanley-Wall NR, Michael AJ. Evolution and multiplicity of arginine decarboxylases in polyamine biosynthesis and essential role in Bacillus subtilis biofilm formation. Journal of Biological Chemistry. 2010;**285**(50):39224-39238

[43] Braz VS, Melchior K, Moreira CG. Escherichia coli as a multifaceted pathogenic and versatile bacterium. Frontiers in Cellular and Infection Microbiology. 2020;**10**:548492

[44] Actor J. Adaptive Immune Response and Hypersensitivity, in Elsevier's Integrated Review Immunology and Microbiology. Philadelphia: Elsevier; 2012

[45] Glasset B, Herbin S, Granier SA, Cavalié L, Lafeuille E, Guérin C, et al. Bacillus cereus, a serious cause of nosocomial infections: Epidemiologic and genetic survey. PLoS One. 2018;**13**(5):0194346

[46] Rodríguez C, Romero E, Garrido-Sanchez L, Alcain-Martinez G, Andrade RJ, Taminiau B, et al. Microbiota insights in Clostridium difficile infection and inflammatory bowel disease. Gut Microbes. 2020;**12**(1):1725220

[47] Aljahdali NH, Sanad YM, Han J, Foley SL. Current knowledge and perspectives of potential impacts of Salmonella enterica on the profile of the gut microbiota. BMC Microbiology. 2020;**20**:1-15

[48] Von Graevenitz A. The role of opportunistic bacteria in human disease. Annual Review of Microbiology. 1977;**31**(1):447-471

[49] Hong HA, Khaneja R, Tam NM, Cazzato A, Tan S, Urdaci M, et al. Bacillus subtilis isolated from the human gastrointestinal tract. Research in Microbiology. 2009;**160**(2):134-143

[50] Schaechter M. Encyclopedia of Microbiology. San Diego, UK: Academic Press; 2009

[51] Gervasi T, Lo Curto R, Minniti E, Narbad A, Mayer MJ. Application of lactobacillus johnsonii expressing phage endolysin for control of Clostridium perfringens. Letters in Applied Microbiology. 2014;**59**(4):355-361

[52] Bartlett JG, Auwaerter PG, Pham PA. Johns Hopkins ABX Guide 2012. Burlington, MA: Jones & Bartlett Publishers; 2011

[53] Dong L, Ariëns RM, America AH, Paul A, Veldkamp T, Mes JJ, et al. Clostridium perfringens suppressing activity in black soldier fly protein preparations. LWT. 2021;**149**:111806

[54] Grenda T, Jarosz A, Sapała M, Grenda A, Patyra E, Kwiatek K. Clostridium perfringens—Opportunistic foodborne pathogen, its diversity and epidemiological significance. Pathogens. 2023;**12**(6):768

[55] Kuwahara T, Yamashita A, Hirakawa H, Nakayama H, Toh H, Okada N, et al. Genomic analysis of Bacteroides fragilis reveals extensive DNA inversions regulating cell surface adaptation. Proceedings of the National Academy of Sciences. 2004;**101**(41):14919-14924

[56] Levinson W. Review of Medical Microbiology and Immunology. New York: McGraw-Hill Education; 2018

[57] Masako K, Hideyuki I, Shigeyuki O, Zenro I. A novel method to control the balance of skin microflora: Part 1. Attack on biofilm of Staphylococcus aureus without antibiotics. Journal of Dermatological Science. 2005;**38**(3):197-205

[58] Banerjee D, Shivapriya P, Gautam PK, Misra K, Sahoo AK, Samanta SK. A review on basic biology of bacterial biofilm infections and their treatments by nanotechnology-based approaches. Proceedings of the National Academy of Sciences, India Section B: Biological Sciences. 2020;**90**:243-259

[59] Donlan RM. Biofilms: Microbial life on surfaces. Emerging Infectious Diseases. 2002;**8**(9):881

[60] Evans LV. Biofilms: Recent Advances in their Study and Control. London: CRC Press; 2000

[61] Dunne WM Jr. Bacterial adhesion: Seen any good biofilms lately? Clinical Microbiology Reviews. 2002;**15**(2):155-166

[62] Cohen BE. Functional linkage between genes that regulate osmotic stress responses and multidrug resistance transporters: Challenges and opportunities for antibiotic discovery. Antimicrobial Agents and Chemotherapy. 2014;**58**(2):640-646

[63] Thoendel M, Kavanaugh JS, Flack CE, Horswill AR. Peptide signaling in the staphylococci. Chemical Reviews. 2011;**111**(1):117-151

[64] Rasamiravaka T, Labtani Q, Duez P, El Jaziri M. The formation of biofilms by Pseudomonas aeruginosa: A review of the natural and synthetic compounds interfering with control mechanisms. BioMed Research International. 2015;**2012**:1-17

[65] Asally M, Kittisopikul M, Rué P, Du Y, Hu Z, Çağatay T, et al. Localized

cell death focuses mechanical forces during 3D patterning in a biofilm. Proceedings of the National Academy of Sciences. 2012;**109**(46):18891-18896

[66] Rathsam C, Eaton RE, Simpson CL, Browne GV, Valova VA, Harty DW, et al. Two-dimensional fluorescence difference gel electrophoretic analysis of Streptococcus m utans biofilms. Journal of Proteome Research. 2005;**4**(6):2161-2173

[67] Islam N, Kim Y, Ross JM, Marten MR. Proteomic analysis of Staphylococcus aureus biofilm cells grown under physiologically relevant fluid shear stress conditions. Proteome Science. 2014;**12**:1-12

[68] Qayyum S, Sharma D, Bisht D, Khan AU. Protein translation machinery holds a key for transition of planktonic cells to biofilm state in enterococcus faecalis: A proteomic approach. Biochemical and Biophysical Research Communications. 2016;**474**(4):652-659

[69] Tielen P, Rosin N, Meyer A-K, Dohnt K, Haddad I, Jänsch L, et al. Regulatory and metabolic networks for the adaptation of Pseudomonas aeruginosa biofilms to urinary tract-like conditions. PLoS One. 2013;**8**(8):71845

[70] Otto M. Staphylococcal infections: Mechanisms of biofilm maturation and detachment as critical determinants of pathogenicity. Annual Review of Medicine. 2013;**64**:175-188

[71] Donlan R. Biofilms: Microbial life on surfaces. Emerging Infectious Diseases. 2002;**9**:881

[72] Al-Wrafy FA, Al-Gheethi AA, Ponnusamy SK, Noman EA, Fattah SA. Nanoparticles approach to eradicate bacterial biofilm-related infections: A critical review. Chemosphere. 2022;**288**:132603

[73] Hall-Stoodley L, Stoodley P. Evolving concepts in biofilm infections. Cellular Microbiology. 2009;**11**(7): 1034-1043

[74] Brooun A, Liu S, Lewis K. A dose-response study of antibiotic resistance in Pseudomonas aeruginosa biofilms. Antimicrobial Agents and Chemotherapy. 2000;**44**(3):640-646

[75] Tseng BS, Zhang W, Harrison JJ, Quach TP, Song JL, Penterman J, et al. The extracellular matrix protects P seudomonas aeruginosa biofilms by limiting the penetration of tobramycin. Environmental Microbiology. 2013;**15**(10):2865-2878

[76] Ciofu O, Tolker-Nielsen T. Tolerance and resistance of Pseudomonas aeruginosa biofilms to antimicrobial agents—How P. Aeruginosa can escape antibiotics. Frontiers in Microbiology. 2019;**10**:913

[77] Vestby LK, Grønseth T, Simm R, Nesse LL. Bacterial biofilm and its role in the pathogenesis of disease. Antibiotics. 2020;**9**(2):59

[78] Remis JP, Costerton JW, Auer M. Biofilms: Structures that may facilitate cell–cell interactions. The ISME Journal. 2010;**4**(9):1085-1087

[79] Zhao G, Usui ML, Lippman SI, James GA, Stewart PS, Fleckman P, et al. Biofilms and inflammation in chronic wounds. Advances in Wound Care. 2013;**2**(7):389-399

[80] Percival SL, McCarty SM, Lipsky B. Biofilms and wounds: An overview of the evidence. Advances in Wound Care. 2015;**4**(7):373-381

[81] Versey Z, da Cruz Nizer WS, Russell E, Zigic S, DeZeeuw KG, Marek JE, et al. Biofilm-innate immune

interface: Contribution to chronic wound formation. Frontiers in Immunology. 2021;**12**:648554

[82] Penesyan A, Paulsen IT, Kjelleberg S, Gillings MR. Three faces of biofilms: A microbial lifestyle, a nascent multicellular organism, and an incubator for diversity. npj Biofilms and Microbiomes. 2021;**7**(1):80

[83] Kerksiek K. A life in slime-biofilms rule the world. Infection Research News and Perspectives. 2008;**9**:1571-1578

[84] Donlan RM, Costerton JW. Biofilms: Survival mechanisms of clinically relevant microorganisms. Clinical Microbiology Reviews. 2002;**15**(2):167-193

[85] Hunt SM, Werner EM, Huang B, Hamilton MA, Stewart PS. Hypothesis for the role of nutrient starvation in biofilm detachment. Applied and Environmental Microbiology. 2004;**70**(12):7418-7425

[86] Hall-Stoodley L, Hu FZ, Gieseke A, Nistico L, Nguyen D, Hayes J, et al. Direct detection of bacterial biofilms on the middle-ear mucosa of children with chronic otitis media. JAMA. 2006;**296**(2):202-211

[87] Elder MJ, Stapleton F, Evans E, Dart JK. Biofilm-related infections in ophthalmology. Eye. 1995;**9**(1):102-109

[88] Johns BE, Purdy KJ, Tucker NP, Maddocks SE. Phenotypic and genotypic characteristics of small colony variants and their role in chronic infection. Microbiology Insights. 2015;**8**:MBI. S25800

[89] Magryś A, Paluch-Oleś J, Bogut A, Kiełbus M, Plewik D, Kozioł-Montewka M. The role of programmed death ligand 1 pathway in persistent biomaterial-associated infections. Journal of Microbiology. 2015;**53**:544-552

[90] Sun J, Deng Z, Yan A. Bacterial multidrug efflux pumps: Mechanisms, physiology and pharmacological exploitations. Biochemical and Biophysical Research Communications. 2014;**453**(2):254-267

[91] Blair JM, Richmond GE, Piddock LJ. Multidrug efflux pumps in gram-negative bacteria and their role in antibiotic resistance. Future Microbiology. 2014;**9**(10):1165-1177

[92] O'Toole G, Kaplan HB, Kolter R. Biofilm formation as microbial development. Annual Reviews in Microbiology. 2000;**54**(1):49-79

[93] Wang L, Slayden RA, Barry CE, Liu J. Cell wall structure of a mutant of mycobacterium smegmatis defective in the biosynthesis of mycolic acids. Journal of Biological Chemistry. 2000;**275**(10):7224-7229

[94] Munita JM, Arias CA. Mechanisms of antibiotic resistance. Virulence Mechanisms of Bacterial Pathogens. 2016;**5**:481-511

[95] Blanco E, Shen H, Ferrari M. Principles of nanoparticle design for overcoming biological barriers to drug delivery. Nature Biotechnology. 2015;**33**(9):941-951

[96] Lee J, Ramasamy M. Recent nanotechnology approaches for prevention and treatment of biofilm-associated infections on medical devices recent nanotechnology approaches for prevention and treatment of biofilm-associated infections on medical devices. BioMed Research International. 2016;**2016**:1851242

[97] Mishra R, Panda AK, De Mandal S, Shakeel M, Bisht SS, Khan J. Natural anti-biofilm agents: Strategies to control biofilm-forming pathogens. Frontiers in Microbiology. 2020;**11**:566325

[98] Bertoglio F, Bloise N, Oriano M, Petrini P, Sprio S, Imbriani M, et al. Treatment of biofilm communities: An update on new tools from the nanosized world. Applied Sciences. 2018;**8**(6):845

[99] Qayyum S, Khan AU. Nanoparticles vs. biofilms: A battle against another paradigm of antibiotic resistance. MedChemComm. 2016;**7**(8):1479-1498

[100] Liu T, Cui Q, Wu Q, Li X, Song K, Ge D, et al. Mechanism study of bacteria killed on nanostructures. The Journal of Physical Chemistry B. 2019;**123**(41):8686-8696

[101] Zwickel T, Kahl SM, Rychlik M, Müller ME. Chemotaxonomy of mycotoxigenic small-spored Alternaria fungi–do multitoxin mixtures act as an indicator for species differentiation? Frontiers in Microbiology. 2018;**9**:1368

[102] Eshed M, Lellouche J, Matalon S, Gedanken A, Banin E. Sonochemical coatings of ZnO and CuO nanoparticles inhibit Streptococcus mutans biofilm formation on teeth model. Langmuir. 2012;**28**(33):12288-12295

[103] Fulaz S, Vitale S, Quinn L, Casey E. Nanoparticle–biofilm interactions: The role of the EPS matrix. Trends in Microbiology. 2019;**27**(11):915-926

[104] Taglietti A, Arciola CR, D'Agostino A, Dacarro G, Montanaro L, Campoccia D, et al. Antibiofilm activity of a monolayer of silver nanoparticles anchored to an amino-silanized glass surface. Biomaterials. 2014;**35**(6):1779-1788

[105] Baptista PV, McCusker MP, Carvalho A, Ferreira DA, Mohan NM, Martins M, et al. Nano-strategies to fight multidrug resistant bacteria—"A Battle of the Titans". Frontiers in Microbiology. 2018;**9**:1441

[106] Vinoj G, Pati R, Sonawane A, Vaseeharan B. In vitro cytotoxic effects of gold nanoparticles coated with functional acyl homoserine lactone lactonase protein from bacillus licheniformis and their antibiofilm activity against Proteus species. Antimicrobial Agents and Chemotherapy. 2015;**59**(2):763-771

[107] Manju S, Malaikozhundan B, Vijayakumar S, Shanthi S, Jaishabanu A, Ekambaram P, et al. Antibacterial, antibiofilm and cytotoxic effects of Nigella sativa essential oil coated gold nanoparticles. Microbial Pathogenesis. 2016;**91**:129-135

[108] Hu D, Li H, Wang B, Ye Z, Lei W, Jia F, et al. Surface-adaptive gold nanoparticles with effective adherence and enhanced photothermal ablation of methicillin-resistant Staphylococcus aureus biofilm. ACS Nano. 2017;**11**(9): 9330-9339

[109] Liu Y, Shi L, Su L, van der Mei HC, Jutte PC, Ren Y, et al. Nanotechnology-based antimicrobials and delivery systems for biofilm-infection control. Chemical Society Reviews. 2019;**48**(2):428-446

[110] Nagvenkar AP, Deokar A, Perelshtein I, Gedanken A. A one-step sonochemical synthesis of stable ZnO–PVA nanocolloid as a potential biocidal agent. Journal of Materials Chemistry B. 2016;**4**(12):2124-2132

[111] Banerjee S, Vishakha K, Das S, Dutta M, Mukherjee D, Mondal J, et al. Antibacterial, anti-biofilm activity and mechanism of action of pancreatin

doped zinc oxide nanoparticles against methicillin resistant Staphylococcus aureus. Colloids and Surfaces B: Biointerfaces. 2020;**190**:110921

[112] Depan D, Misra R. On the determining role of network structure titania in silicone against bacterial colonization: Mechanism and disruption of biofilm. Materials Science and Engineering: C. 2014;**34**:221-228

[113] Wang Z, Dong K, Liu Z, Zhang Y, Chen Z, Sun H, et al. Activation of biologically relevant levels of reactive oxygen species by Au/g-C3N4 hybrid nanozyme for bacteria killing and wound disinfection. Biomaterials. 2017;**113**:145-157

[114] Garza-Cervantes JA, Escárcega-González CE, Barriga Castro ED, Mendiola-Garza G, Marichal-Cancino BA, López-Vázquez MA, et al. Antimicrobial and antibiofilm activity of biopolymer-Ni, Zn nanoparticle biocomposites synthesized using R. Mucilaginosa UANL-001L exopolysaccharide as a capping agent. International Journal of Nanomedicine. 2019;**14**:2557-2571

[115] Trigo Gutierrez JK, Zanatta GC, Ortega ALM, Balastegui MIC, Sanita PV, Pavarina AC, et al. Encapsulation of curcumin in polymeric nanoparticles for antimicrobial photodynamic therapy. PLoS One. 2017;**12**(11):0187418

[116] Baek J-S, Tan CH, Ng NKJ, Yeo YP, Rice SA, Loo SCJ. A programmable lipid-polymer hybrid nanoparticle system for localized, sustained antibiotic delivery to gram-positive and gram-negative bacterial biofilms. Nanoscale Horizons. 2018;**3**(3):305-311

[117] Ramos MADS, Da Silva PB, Spósito L, De Toledo LG, Bonifacio BV, Rodero CF, et al. Nanotechnology-based drug delivery systems for control of microbial biofilms: A review. International Journal of Nanomedicine. 2018;**13**:1179

[118] Mei L, Xu Z, Shi Y, Lin C, Jiao S, Zhang L, et al. Multivalent and synergistic chitosan oligosaccharide-Ag nanocomposites for therapy of bacterial infection. Scientific Reports. 2020;**10**(1):10011

[119] Khan F, Manivasagan P, Pham DTN, Oh J, Kim S-K, Kim Y-M. Antibiofilm and antivirulence properties of chitosan-polypyrrole nanocomposites to Pseudomonas aeruginosa. Microbial Pathogenesis. 2019;**128**:363-373

[120] Horev B, Klein MI, Hwang G, Li Y, Kim D, Koo H, et al. pH-activated nanoparticles for controlled topical delivery of farnesol to disrupt oral biofilm virulence. ACS Nano. 2015;**9**(3):2390-2404

[121] Sims KR, Liu Y, Hwang G, Jung HI, Koo H, Benoit DS. Enhanced design and formulation of nanoparticles for anti-biofilm drug delivery. Nanoscale. 2019;**11**(1):219-236

[122] Pircalabioru GG, Chifiriuc M-C. Nanoparticulate drug-delivery systems for fighting microbial biofilms: From bench to bedside. Future Microbiology. 2020;**15**(8):679-698

Chapter 3

Bacterial Biofilm Eradication in Human Infections

Chin Erick Ngehdzeka and Zeuko'o Menkem Elisabeth

Abstract

Microbial biofilms are elaborate and highly resistant aggregates formed on surfaces or medical devices, causing two-thirds of infections and leading to a serious threat to public health. Their presence increases the rate of infections and mortality in the affected individuals. The strategies and eradication patterns are necessary to be established or implemented to eliminate them in human beings. This chapter highlights recent approaches for combating bacterial biofilms, including the methods used by promising antibiofilm compounds to enhance the total elimination of bacterial biofilms involved in some specific human infections. Biofilms must be eradicated to ensure efficient treatment of the infections.

Keywords: bacterial biofilms, eradication, human infections, resistance, aggregates

1. Introduction

Bacterial drug resistance and biofilm infections can result in a wide range of diseases and associated complications, such as sepsis, endocarditis, pneumonia, and even death in the worst scenarios [1, 2]. Bacterial biofilms are complex and elaborate microbial communities that are very resistant and readily colonizing the surfaces of organs or medical implants to cause intractable and recurring infections [3]. They have a large spectrum of activities ranging from nosocomial setting, especially linked to lower respiratory, urinary tract, and surgical wound infections as well as the medical devices used during treatment resulting to a serious challenge to patients' health [4]. Bacteria tend to work in synergy and create groups to achieve resistance about 10–1000 times on antibiotics and the human immune system [3, 4], while also secreting various virulence factors in certain cases [2]. The arrangement of bacteria in the biofilm in a micro-colony shape enclosed in an extracellular polymeric substance (EPS) of the matrix, is the highest surviving mechanisms of biofilms giving them more resilience and versatility [5]. Fleming's discovery of penicillin in 1928 marked the advent of antibiotics and the subsequent production of several antibiotics has been a lifesaving against bacterial infections [6]. Many strategies are being used today to control and eradicate important biofilms that ranges from drugs and cell methods to non-biological modern technologies. These include novel antibiotics and their carriers, bacteriophage and its components, antiseptics and disinfectants, small molecule anti-biofilm agents, surface treatment strategies, ultrasound-induced microbubbles,

IntechOpen

nanomaterials and nanostructure functionalization, as well as multifunctional coating [4, 7, 8]. This chapter therefore, highlights recent approaches for combating bacterial biofilms, including methods used by promising antibiofilm compounds to enhance the total elimination of bacterial biofilms involved in some specific human infections.

1.1 Bacterial biofilms in human infections: Examples and consequences

Bacteria can grow to a biofilm during favorable condition. Notwithstanding certain species seem to have a preference to form biofilms and examples of these are given in **Table 1** [9]. Centuries ago it was thought that bacteria only existed in free floating forms or as planktonic organisms until the 1970s when they were observed adhering and growing on surfaces [10].

Most of these species form biofilms at their natural sites and constitute the microflora in the human body (**Figure 1**) [9]. Biofilms are extremely resistant to orthodox antimicrobial treatment and to the host immune response. It is reported to play a key role in various chronic infections in human diseases, thereby representing a challenge in clinical settings [9, 11]. The control and treatment of infections caused by biofilms are challenging in medical settings and this has led to the development of novel technology and new strategies to combat microbial biofilms [9].

1.2 Bacterial biofilm eradication mechanisms

Biofilms are difficult to eradicate even with extended treatment and important attention is given to less conventional treatments, especially those that employ macromolecular species [12, 13]. Because of this failure of conventional therapeutics biofilms require other strategies and mechanism for their elimination [14]. Many strategies are presently under investigation, aiming to effectively eradicate biofilm-related infections and several agents have been known to have anti-biofilm activity, including some natural products, synthetic compounds, enzymes, peptides, chelating agents, polyphenols, as well as some antibiotics [13, 14]. These biofilms have different mechanisms of action.

Organism	Site of biofilm formation
Actinomyces spp.	Teeth
Escherichia coli and other enterobacteria	Urinary catheters
Escherichia coli	Intestinal tract
Lactobacillus spp.	Vagina, Teeth
Pseudomonas aeruginosa	Lungs of cystic fibrosis patients
Staphylococcus aureus	Implantable medical devices
Staphylococcus epidermidis and other coagulase-negative staphylococci	Implantable medical devices
Streptococcus spp.	Teeth

Table 1.
Examples of representative bacterial pathogens that frequently form biofilms.

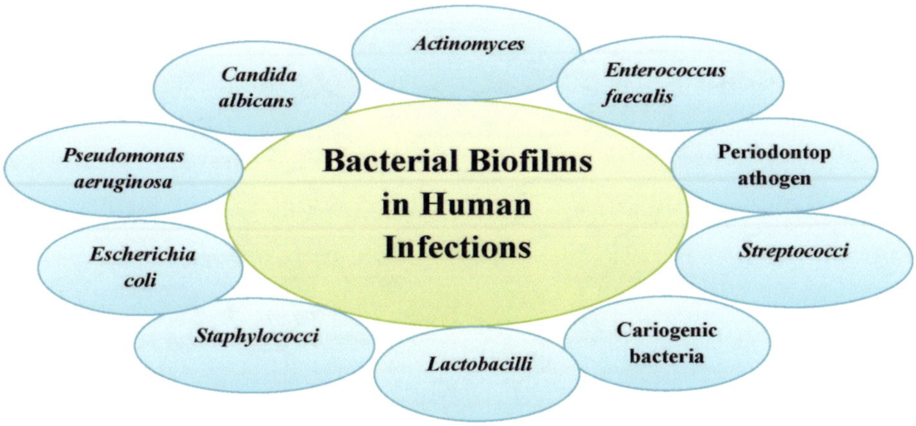

Figure 1.
Bacterial biofilm in human infections.

2. N-acyl homoserine lactones mediated quorum sensing inhibition

Quorum sensing (QS) is a system where bacterial cells communicate through the activation of specific signals, with key objective of enabling the adaptation of bacteria hostile environmental conditions, including bacterial population densities. This process involves reacting to extracellular chemical signaling molecules called autoinducers (AIs) through synthetization and sensing. Gram-negative bacteria communicate using AIs, most commonly acyl-homoserine lactones (AHLs) and other small molecules [14]. The mechanism is the disruption of AIs and then mitigate quorum sensing controlled responses for biofilm control. Most of the anti-biofilm chemical structures under studies are: N-acyl homoserine lactones (AHL) (**Figure 2a**), triazole dihydro furanone (**Figure 2b**), synthetic halogenated furanone (**Figure 2c**), EGCG (**Figure 2d**), and ellagic acid (**Figure 2e**) [14]. Numerous AHLs disrupt biofilm formation. Important biofilm inhibitory effect against *P. aeruginosa* and *Serratia marcescens* were observed when the lactone moiety of the native AHL molecules is replaced by cyclohexanone or cyclopentyl [15, 16].

3. Membrane permeabilization and potential alteration

Pore formation and destruction of the cytoplasmic membrane is as a result of bacterial membrane modification. There are three possible mechanisms of bacterial membrane disruption by antimicrobial peptides (AMPs): (a) pore-induced barrel-stave pathway, (b) toroidal pathway, and (c) carpet (non-pore) mode (**Figure 3**) [17]. Peptides that inhibit bacteria by disrupting their membranes and consequently inhibiting enzyme production are produced and post-translationally modified. These peptides are lantibiotics that are ring-structured peptide antibiotics containing thioether amino acids (methyllanthionine or lanthionine) or unsaturated amino acids (2-amino isobutyric acids or dehydro-alanine) [18]. A pore-forming lantibiotic called subtilin, produced from a Gram-positive bacteria *B. subtilis* strain ATCC6633, induces the dissipation of transmembrane electrostatic-potential releasing cytoplasmic solutes from *B. subtilis* and *Staphylococcus simulans* membrane vesicles [19]. In **Figure 2**, AMP outreaches the cytoplasmic membrane via permeabilizing the outer membrane in Gram-negative bacteria,

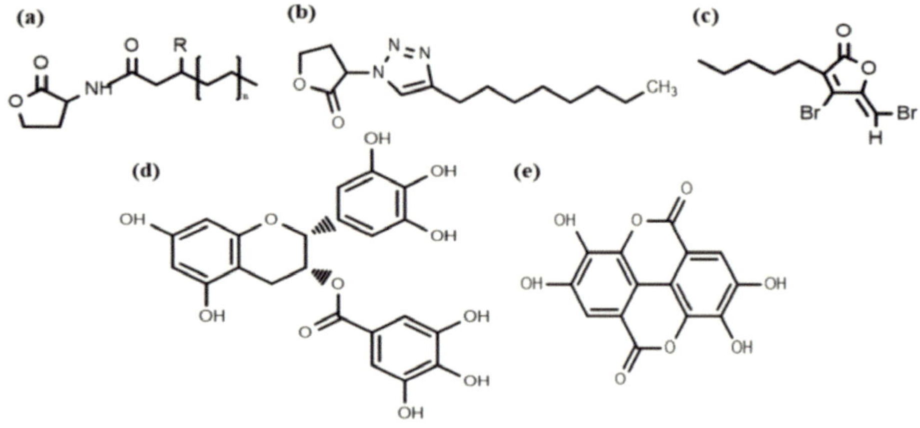

Figure 2.
Chemical structures of some anti-biofilm compounds that inhibit AHL-mediated QS. (a) AHL. (b) Triazole dihydro furanone. (c) Synthetic halogenated furanone. (d) Epigallocatechin gallate (EGCG). (e) Ellagic acid.

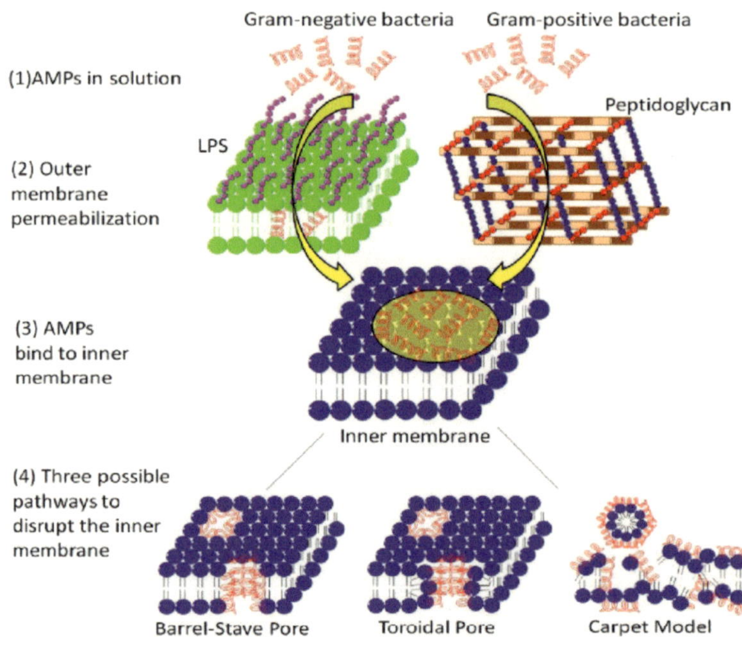

Figure 3.
Mechanism of action of AMPs on the membrane system of Gram-negative and Gram-positive bacteria [17].

while in Gram-positive bacteria, the AMP directly disperses through nano-ranged pores of the peptidoglycan layer. After binding to the inner membrane, AMPs can create three types of pores, that is, barrel-stave pore, toroidal pore, and carpet model [17].

4. Peptidoglycan cleavage

Peptidoglycan, the cleavage of which is also known to inhibit biofilm formation, is a layer located in the cell walls of many bacteria and originates from amino acids and sugars [20]. Peptidoglycan cleavage causes a change in protein composition and amount of teichoic acid in the bacterial cell wall resulting to biofilm inhibition [20]. An example of peptidoglycan hydrolases is endolysin encoded by bacteriophages [21]. Endolysin can work on multidrug-resistant strains, by disrupting biofilms *in vitro* e.g., PlyC (specific Streptococcal bacteriophage) [22].

5. Inhibition of bacterial cell division

One mechanism used to stop biofilms from growing is to inhibit cell division. Peptides having antimicrobial activity inhibit cytoplasmic proteins that play a big role in cell division and also promote cell growth by penetrating the bacterial cytosol through formation of channels at outer membrane or via a flip-flop of phospholipids (when the cell is ready to divide, then the nuclear membrane melts) [14]. Drosocin, pyrrhocoricin, and apidaecin are proline-rich antimicrobial peptides (AMPs). They can impede the initiation of chromosomal DNA (cDNA) replication by binding with a shock protein of bacteria DnaK [23, 24] or bacterial death [25].

6. Biofilm's inhibitors based on nucleotide second messenger molecules

Nucleotide second messenger molecule cyclic di-GMP (c-di-GMP) is involved in biofilm development and the growth of biofilm can be altered by modifying the c-di-GMP signaling pathway **(Figure 4)** [14, 26]. c-di-GMP is synthesized from two molecules of GTP by diguanylate cyclases (DGCs). Its mechanism of action is achieved by microbial cells reducing the level of c-di-GMP via phosphodiesterase activation due to nitrosative and starvation conditions [27] leading to biofilm dispersion. However, c-di-GMP has three main mechanisms of biofilm formation regulation:

a. Weakening bacterial movement to promote bacterial attachment onto a solid surface. Transition of bacteria from motility to attachment is an important stage in biofilm formation. In *E. coli*, the c-di-GMP-bound form of the flagellar brake protein YcgR interacts with the flagellar motor protein Mot A, thus regulatory motor output in a brake-like fashion [28]. Bacterial surface attachment is enhanced by, c-di-GMP with a role in the suppression of bacteria motility **(Figure 4A)**.

b. Regulating pilus development. Mannose-sensitive haemagglutinin (MsHA) pilus found in *Vibrio cholera* promotes bacteria attachment on solid surfaces during the early stages of biofilm formation and controlled by MshE, an ATPase responsible for pilus polymerization [29, 30]. Once c-di-GMP binds to MshE, it enhances the assembly of MsHA pilus and thereby increases biofilm formation [29] **(Figure 4B).** The number of MsHA pili on the bacterial surface can also increase proportionally with increase in intracellular c-di-GMP concentration, leading to rapid biofilm formation.

c. Regulation of biofilm components production. Curli fibers is regulated by the diguanylate cyclases (DGC YdaM) and the phosphodiesterases (PDE YciR) through controlling the c-di-GMP concentrations in *E. coli* K-12 strain W3110, production of biofilm matrix components [31]. At very high concentrations

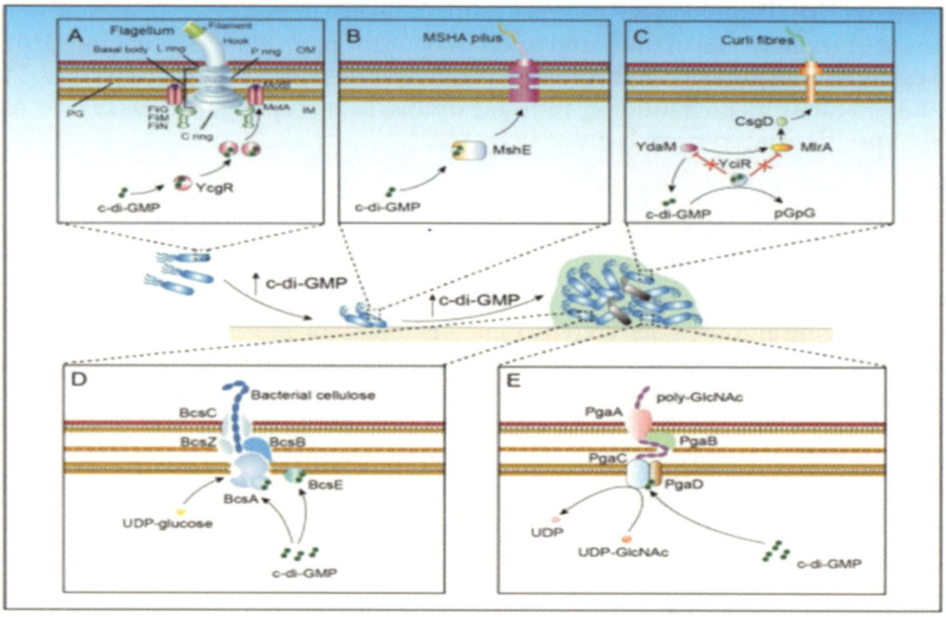

Figure 4.
Regulation of cyclic di-GMP on biofilm formation by inhibiting bacterial motility and increasing EPS production [26].

of c-di-GMP, YciR begins to apply its PDE function to release the inhibition of YdaM and MlrA and concomitantly, YdaM can activate MlrA to enhance the central curli regulator CsgD, thus prompting the transcription of curli genes and enabling the curli formation [31] (**Figure 4C**). c-di-GMP also regulates another essential component of biofilm matrix called bacteria cellulose [32]. Bacterial cellulose synthase (BcsA), is attached in the inner membrane of the cell and contains a catalytic glycosyltransferase domain and a c-di-GMP-binding PilZ domain in its intracellular part [33] (**Figure 4D**). Glycosyltransferase domain is activated when c-di-GMP binds to the PilZ domain of BcsA which allows the bacterial cell to assemble the nascent polysaccharide with the help of the BcsB/BcsC/BcsZ complex to form extracellular cellulose [33] (**Figure 4D**).

7. Application of external pressures to eradicate biofilm

Several biochemical and physical methods can be used to eradicate formed biofilms as described in **Figure 5.**

a. Physical methods: This include treatment with ultrasound and magnetic fields as shown in **Figure 5A** [34]. The forces produced by cavitation bubbles and fluid movement are primarily what cause biofilm dispersion during ultrasonic biofilm removal.

b. Biochemical methods: Application of phage lysins, degradative enzymes, and microbial metabolites (**Figure 5B**). Humans' enemies are pathogenic bacteria, and bacteria's enemies are bacteriophages. Phage lysins, the weapons of bacteriophages, could be used to combat detrimental biofilms and multidrug-resistant microorganisms (**Figure 5B**). At a later stage of infection, bacteriophages express peptidoglycan hydrolases known as phage lysins. They have the ability to destroy bacteria by cleaving the peptidoglycan layer of the cell wall (**Figure 5B**) [35].

c. Degradative enzymes: Through the degradation of EPS and removing this protective clothing (**Figure 5C**) [26]. Exopolysaccharides, proteins, lipids, and nucleic acids make up the majority of the biofilm matrix in EPS. Degrading the EPS and eliminating the "protective clothing" that biofilm provides for microorganisms allows one to get rid of the hazardous biofilm. Emerging nosocomial pathogen *Corynebacterium auris* has a mannan-glucan-rich matrix in its biofilms. Infections brought on by this organism were successfully treated by hydrolyzing mannan-glucan in the biofilm matrix using mannosidase or glucanase. Additionally, cystic fibrosis patients' sputum and human lung tissue have both been found to have *Pseudomonas* cells encased in alginate-rich biofilms. By eliminating exopolysaccharide from the *P. aeruginosa* cell surface and converting alginate into unsaturated uronic acid-containing oligosaccharides, alginate lyase has been used to aid in the removal of biofilms [26].

d. Microbial metabolites: Various microbial physiological processes, including the development of biofilms, have been revealed to be regulated by secondary metabolites acting as intercellular signals. As a result, metabolites can also be used to regulate the growth of biofilms (**Figure 5D**) [36]. By generating cell chain elongation, morphological alterations, and even cell death in *S. mutans* biofilms, carolacton, a secondary metabolite isolated from *Sorangium cellulosum*, shown excellent eradication activity against *S. mutans* biofilms. The biosurfactant made by Pseudomonas spp., rhamnolipid, demonstrated the capacity to dislodge and eliminate *S. aureus* biofilms.

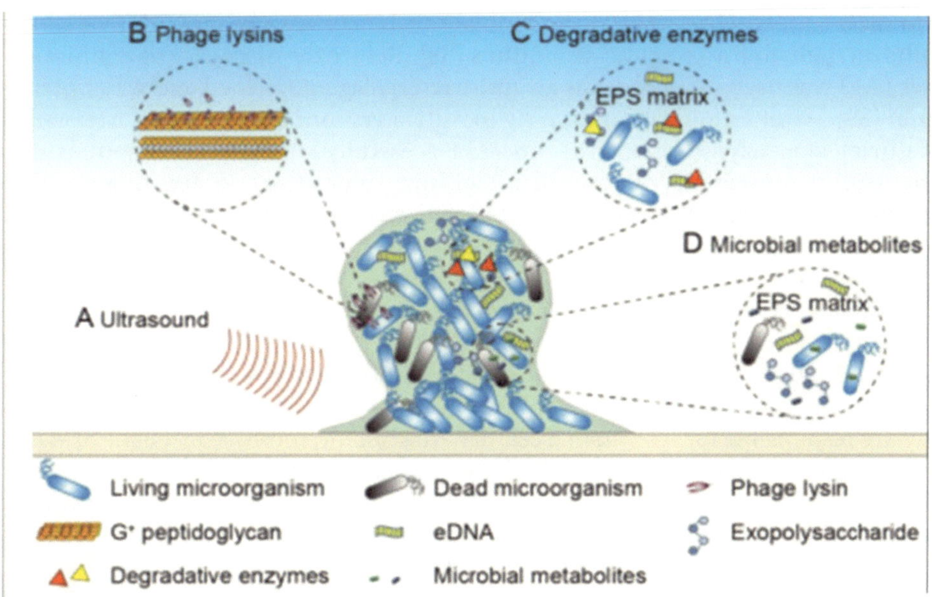

Figure 5.
Application of external pressures to eradicate mature biofilm, which include (A) ultrasound, (B) phage lysins, (C) degradative enzymes, and (D) microbial metabolites [26].

e. Nitric oxide: At low and non-toxic concentrations, NO produced by the anaerobic respiration activities inside the *P. aeruginosa* biofilm can start the dispersal of the biofilm. Further investigation indicated that *P. aeruginosa* biofilm NO signaling can increase PDE activity, lowering intracellular c-di-GMP levels and promoting biofilm dispersion. NO-induced biofilm dispersal was also seen in several other bacteria, including *E. coli* and *S. aureus*, in addition to *P. aeruginosa*. Similarly, exogenous NO addition therapy can promote biofilm dispersion. For instance, NO-releasing polymers have the ability to dose-dependently lower the metabolic activity of different biofilms. Additionally, regardless of the matrix's composition, NO-releasing cyclodextrins can destroy *P. aeruginosa* biofilm. A supramolecular nanocarrier was created by combining the NO prodrug with the glutathione-sensitive a-cyclodextrin and chlorin e6 prodrug demonstrate quick NO release when glutathione is overexpressed in the biofilm, effectively destroying the *S. aureus* biofilm [37].

7.1 Biofilm eradication novel perspectives

Bacteria resistance to antibiotics is fast expanding and orthodox treatment of biofilms with antibiotics is ineffective. Combating biofilms therefore requires the development of different approaches and biologists looking for alternative ways to eradicate biofilms. Some recent approaches for combating biofilms include electrochemical methods, promising antibiofilm compounds and drug delivery strategies to enhance the bioavailability and potency of antibiofilm agents [38].

7.1.1 Antimicrobial compounds to eradicate biofilms: Substrates with antibiofilm activity

In recent approaches many antimicrobial substances are screened with variable results on their ability to eradicate biofilms [38]. Before the discovery of antibiotics, silver (Ag) was used for decades as an anti-bacterial agent for food and water preservation [39]. After its ionization from Ag to Ag^+, silver ions are capable of irreversibly denaturing key enzymes and in the process successfully killing mature biofilms after 24 hours of treatment [39, 40] as well as biofilms grown after 4–6 days though higher concentrations are needed. Biofilms that have been successfully eradicated by silver oxynitrate include *E. coli*, *S. aureus,* and *P. aeruginosa* biofilms [40].

7.1.2 Modulation of the biofilm architecture to eradicate biofilms

Biofilm extracellular polymeric substances (EPS) called the matrix is composed of proteins, polysaccharides, and eDNA and loosely links bacteria within the biofilm. They are responsible for irreversible cellular attachment, improve mechanical stability, and maintain secreted enzymes [41]. Theoretically, biofilm EPS matrix targeted agents have the possibility to interfere with the growth of biofilm, their dislocation, destabilization, detachment, sensitization, and increase access of antibiotics [38]. An example is the inhibition of biofilm formation by a variety of Gram-positive and Gram-negative organisms, e.g., *S. epidermidis and P. aeruginosa*. Specifically, deoxyribonuclease I (DNase I) cleaves single-stranded or double-stranded DNA at phosphodiester bonds that make up the phosphate backbone during the addition of DNase I [41–43].

7.1.3 Electrochemical biofilm eradication

One of the new perspectives in eradicating biofilms is through electrochemical techniques. The mechanism of action involved is the formation of H_2O_2, which is a result of the partial reduction of oxygen on metal surfaces [44]. This stimulates an electric current that affects the organization of biological membranes, cellular processes [45], cell behavior [46], bacterial respiratory rate, and oxidation of proteins, likewise cell electrophysiology [47]. The eradication of biofilm using electrochemically generated biocides varies depending on biocide concentration, exposure time, biofilm thickness and/or growth stage, and bacterial strain, as observed [44]. An example of experiment carried out *in vivo* on *Acinetobacter baumannii* grown as biofilms on porcine explants showed it could be overlaid with the same e-scaffold, and this significantly reduced viable bacteria by about 1000 folds [48].

7.2 Strategies for combating bacterial biofilm infections

Strategies to combat biofilm formation range from the control by surface adhesins to the control by cell-to-cell communication pathways [49]. Three strategies have been identified, which include (a) altering abiotic surface characteristics to prevent biofilm formation; (b) regulating the signaling pathways to inhibit biofilm formation and stimulate biofilm dispersal; and (c) applying external forces to eradicate the biofilm (**Figure 6**) [26].

7.2.1 Altering abiotic surface characteristics to prevent biofilm formation

Here two strategies are characterized, that is, treating abiotic surfaces and coating surfaces (**Figure 6A–B**) [26]. Treatment relies on changing the characteristics of surface material like smoothness and wettability through thermal cycling and UV irradiation or hydrophilicity [50] and coating surfaces with polymers, trimethyl-silane (TMS)/O_2, and antimicrobial peptides in order to prevent biofilm attachment [26].

7.2.2 Regulation of signaling pathways to inhibit biofilm

Examples of pathways inhibition are those based on quorum sensing that triggers a cascade of intracellular signaling events, eventually regulating different physiological phenotypes after binding to their matching receptors [51], inhibiting biofilm-related genes expression through the interfering with the QS signaling pathway [26], and inhibition based on nucleotides (**Figure 6C**) [27].

7.2.3 Application external forces to eradicate the biofilm

Both physical and biochemical methods are used to eradicate already formed biofilm. Physical methods entail use of UV radiation whereas biochemical methods use phage lysins, degradative enzymes, metabolites and nitric oxides (**Figure 6D**) [26].

7.3 Treatment methods for biofilm infections

Biofilm infections can be reduced by using conventional antibiotics, either alone or in combination with additional medicines (**Figure 7**). For example, it was observed that sub-minimum inhibitory concentrations (MICs) of ceftazidime repressed the

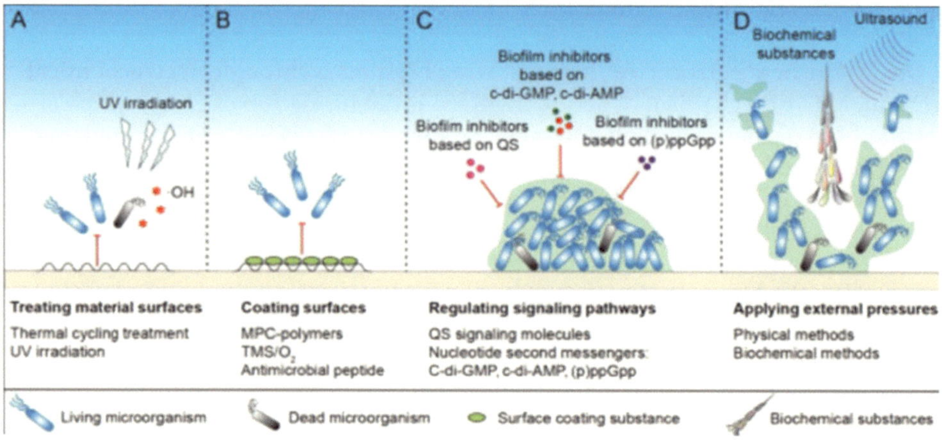

Figure 6.
Strategies for controlling biofilm infections. (A) Surface treatment. (B) Surface coating. (C) Chemical agents that influence QS. (D) External force [26].

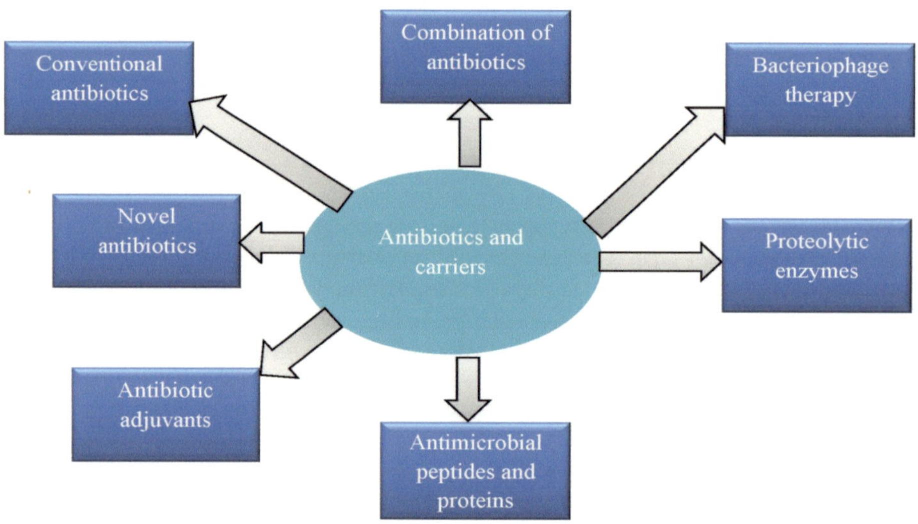

Figure 7.
Schematic illustration of antibiotics-based tactics for preventing the growth of clinically-significant bacterial biofilms [7].

expression of genes involved in *P. aeruginosa* bacterial adherence and matrix synthesis, decrease biofilm volume, and impede twitching motility [7, 52]. Colistin also greatly decreased *E. coli* biofilms and planktonic cells in a concentration-dependent manner [53]. According to an *in vitro* study, gentamycin released by bone graft replacements can inhibit *E. coli* adhesion at 12 μg/mL and can remove biofilms that have been present for 24 hours at 23 μg/mL [54].

However, because biofilm has emerged, the majority of antibiotics are now given in clinical settings in mixtures with other antibiotics. Despite the fact that vancomycin is still the antibiotic most frequently recommended for *S. aureus* biofilm-associated infections, the rise of vancomycin-resistant *S. aureus* has made it necessary to combine vancomycin with other antibiotics, such as rifampin. Additionally, colistin and other antibiotics, such tigecycline, have demonstrated synergistic effects *in vitro*, indicating the possibility of their use in clinical settings. Also, it was proven that various strains of *E. coli* linked to UTIs had their biofilm biomass drastically decreased by amikacin, ciprofloxacin, and third-generation cephalosporins. Additionally, it was shown that *Staphylococcal* biofilms developed on titanium devices may be removed within 72 hours with the help of a combined antibiotic therapy of clarithromycin and daptomycin [7].

8. Conclusion

Complex and dynamic interactions between the surface, microorganisms, and EPS are necessary for the development of a biofilm. In addition, biofilms contain a form of bacteria that is common in nature. Their resistance is a major barrier that traditional methods must overcome. However, antimicrobials have not received enough attention at current stage. The spatial heterogeneity in the chemical and microbial composition of biofilms has made it more challenging to execute eradication strategies. For the purpose of preventing infections, this paper highlighted a number of cutting-edge antimicrobials based on nanotechnology and delivery techniques, especially in the context of better penetration and targeted antimicrobial administration inside the biofilm for its eradication.

Author details

Chin Erick Ngehdzeka[1,2] and Zeuko'o Menkem Elisabeth[3,4]*

1 Faculty of Medicine and Biomedical Sciences, Department of Public Health, University of Yaoundé 1, Yaounde, Cameroon

2 St. Pietro Health Center, Yaounde, Cameroon

3 Faculty of Health Sciences, Department of Biomedical Sciences, University of Buea, Buea, Cameroon

4 Faculty of Science, Antimicrobial and Biocontrol Agents Unit, Laboratory for Phytobiochemistry and Medicinal Plant Studies, Department of Biochemistry, University of Yaoundé I, Yaoundé, Cameroon

*Address all correspondence to: zeukoo@yahoo.com

References

[1] Ouyang J, Bu Q, Tao N, Chen M, Liu H, Zhou J, et al. A facile and general method for synthesis of antibiotic-free protein-based hydrogel: Wound dressing for the eradication of drug-resistant bacteria and biofilms. Bioactive Materials. 2022;**18**:446-458

[2] Tan X, Xie S, Jin X, Zhu R, Wei S, Song Y, et al. Four temporin-derived peptides exhibit antimicrobial and antibiofilm activities against methicillin-resistant Staphylococcus aureus. Acta Biochimica et Biophysica Sinica [Internet]. 2022. Available from: https://engine.scichina.com/doi/10.3724/abbs.2022013 [Accessed: March 4, 2023]

[3] Dong Y, Wang L, Zhang Z, Ji F, Chan TKF, Yang H, et al. Endoscope-assisted magnetic helical micromachine delivery for biofilm eradication in tympanostomy tube. Science Advances. 2022;**8**(40):eabq8573

[4] Chang C, Yu X, Guo W, Guo C, Guo X, Li Q, et al. Bacteriophage-mediated control of biofilm: A promising new dawn for the future. Frontiers in Microbiology [Internet]. 2022;**13**. Available from: https://www.frontiersin.org/articles/10.3389/fmicb.2022.825828 [Accessed: March 10, 2023]

[5] Bi Y, Xia G, Shi C, Wan J, Liu L, Chen Y, et al. Therapeutic strategies against bacterial biofilms. Fundamental Research. 2021;**1**(2):193-212

[6] Crofts TS, Gasparrini AJ, Dantas G. Next-generation approaches to understand and combat the antibiotic resistome. Nature Reviews Microbiology. 2017;**15**(7):422-434

[7] Krukiewicz K, Kazek-Kęsik A, Brzychczy-Włoch M, Łos MJ, Ateba C, Mehrbod P, et al. Recent advances in the control of clinically important biofilms. International Journal of Molecular Sciences. 2022;**23**(17):9526

[8] Darvishi S, Tavakoli S, Kharaziha M, Girault HH, Kaminski CF, Mela I. Advances in the sensing and treatment of wound biofilms. Angewandte Chemie International Edition. 2022;**61**(13):e202112218

[9] Wilson M. Bacterial biofilms and human disease. Science Progress. 2001;**84**(3):235-254

[10] Yadav MK, Song JJ, Singh BP, Vidal JE. Microbial biofilms and human disease: A concise review. In: New and Future Developments in Microbial Biotechnology and Bioengineering: Microbial Biofilms [Internet]. Elsevier; 2020. pp. 1-13. Available from: https://linkinghub.elsevier.com/retrieve/pii/B9780444642790000013 [Accessed: March 14, 2023]

[11] Sabir N, Ikram A, Zaman G, Satti L, Gardezi A, Ahmed A, et al. Bacterial biofilm-based catheter-associated urinary tract infections: Causative pathogens and antibiotic resistance. American Journal of Infection Control. 2017;**45**(10):1101-1105

[12] Blackman LD, Qu Y, Cass P, Locock KES. Approaches for the inhibition and elimination of microbial biofilms using macromolecular agents. Chemical Society Reviews. 2021;**50**(3):1587-1616

[13] Luo Y, Yang Q, Zhang D, Yan W. Mechanisms and control strategies of antibiotic resistance in pathological biofilms. Korean Society for

Microbiology and Biotechnology. 28 Jan 2021;**31**(1):1-7

[14] Asma ST, Imre K, Morar A, Herman V, Acaroz U, Mukhtar H, et al. An overview of biofilm formation–combating strategies and mechanisms of action of antibiofilm agents. Life. 2022;**12**(8):1110

[15] Ishida T, Ikeda T, Takiguchi N, Kuroda A, Ohtake H, Kato J. Inhibition of quorum sensing in Pseudomonas aeruginosa by N-acyl cyclopentylamides. Applied and Environmental Microbiology. 2007;**73**(10):3183-3188

[16] Morohoshi T, Shiono T, Takidouchi K, Kato M, Kato N, Kato J, et al. Inhibition of quorum sensing in Serratia marcescens AS-1 by synthetic analogs of N-acylhomoserine lactone. Applied and Environmental Microbiology. 2007;**73**(20):6339-6344

[17] Li J, Koh JJ, Liu S, Lakshminarayanan R, Verma CS, Beuerman RW. Membrane active antimicrobial peptides: Translating mechanistic insights to design. Frontiers in Neuroscience [Internet]. 2017;**11**. Available from: https://www.frontiersin.org/articles/10.3389/fnins.2017.00073 [Accessed: March 18, 2023]

[18] Bierbaum G, Sahl HG. Lantibiotics: Mode of action, biosynthesis and bioengineering. Current Pharmaceutical Biotechnology. 2009;**10**(1):2-18

[19] Parisot J, Carey S, Breukink E, Chan WC, Narbad A, Bonev B. Molecular mechanism of target recognition by subtilin, a class I lanthionine antibiotic. Antimicrobial Agents and Chemotherapy. 2008;**52**(2):612-618

[20] Shah IM, Laaberki MH, Popham DL, Dworkin J. A eukaryotic-like Ser/Thr kinase signals bacteria to exit dormancy in response to peptidoglycan fragments. Cell. 2008;**135**(3):486-496

[21] Shen Y, Köller T, Kreikemeyer B, Nelson DC. Rapid degradation of Streptococcus pyogenes biofilms by PlyC, a bacteriophage-encoded endolysin. The Journal of Antimicrobial Chemotherapy. 2013;**68**(8):1818-1824

[22] Hoopes JT, Stark CJ, Kim HA, Sussman DJ, Donovan DM, Nelson DC. A novel use of a bacteriophage lysin, PlyC, as a disinfectant against Streptococcus equi. Applied and Environmental Microbiology. 2009;**75**(5):1388-1394

[23] Kragol G, Hoffmann R, Chattergoon MA, Lovas S, Cudic M, Bulet P, et al. Identification of crucial residues for the antibacterial activity of the proline-rich peptide, pyrrhocoricin: Multifunctional antimicrobial peptides. European Journal of Biochemistry. 2002;**269**(17):4226-4237

[24] Kragol G, Lovas S, Varadi G, Condie BA, Hoffmann R, Otvos L. The antibacterial peptide pyrrhocoricin inhibits the ATPase actions of DnaK and prevents chaperone-assisted protein folding. Biochemistry. 2001;**40**(10):3016-3026

[25] Gagnon MG, Roy RN, Lomakin IB, Florin T, Mankin AS, Steitz TA. Structures of proline-rich peptides bound to the ribosome reveal a common mechanism of protein synthesis inhibition. Nucleic Acids Research. 2016;**44**(5):2439-2450

[26] Yin W, Xu S, Wang Y, Zhang Y, Chou SH, Galperin MY, et al. Ways to control harmful biofilms: Prevention, inhibition, and eradication. Critical Reviews in Microbiology. 2021;**47**(1):57-78

[27] Römling U, Galperin MY, Gomelsky M. Cyclic di-GMP: The first 25 years of a universal bacterial second messenger. Microbiology and Molecular Biology Reviews: MMBR. 2013;**77**(1):1-52

[28] Hou YJ, Yang WS, Hong Y, Zhang Y, Wang DC, Li DF. Structural insights into the mechanism of c-di-GMP-bound YcgR regulating flagellar motility in Escherichia coli. The Journal of Biological Chemistry. 2020;**295**(3):808-821

[29] Jones CJ, Utada A, Davis KR, Thongsomboon W, Zamorano Sanchez D, Banakar V, et al. C-di-GMP regulates motile to sessile transition by modulating MshA pili biogenesis and near-surface motility behavior in Vibrio cholerae. Parsek MR, editor. PLoS Pathogens. 2015;**11**(10):e1005068

[30] Wang YC, Chin KH, Tu ZL, He J, Jones CJ, Sanchez DZ, et al. Nucleotide binding by the widespread high-affinity cyclic di-GMP receptor MshEN domain. Nature Communications. 2016;**7**:12481

[31] Lindenberg S, Klauck G, Pesavento C, Klauck E, Hengge R. The EAL domain protein YciR acts as a trigger enzyme in a c-di-GMP signalling cascade in E. coli biofilm control. The EMBO Journal. 2013;**32**(14):2001-2014

[32] Römling U, Galperin MY. Bacterial cellulose biosynthesis: Diversity of operons, subunits, products, and functions. Trends in Microbiology. 2015;**23**(9):545-557

[33] Morgan JLW, Strumillo J, Zimmer J. Crystallographic snapshot of cellulose synthesis and membrane translocation. Nature. 2013;**493**(7431):181-186

[34] Vyas N, Wang QX, Manmi KA, Sammons RL, Kuehne SA, Walmsley AD. How does ultrasonic cavitation remove dental bacterial biofilm? Ultrasonics Sonochemistry. 2020;**67**:105112

[35] Drulis-Kawa Z, Majkowska-Skrobek G, Maciejewska B. Bacteriophages and phage-derived proteins--application approaches. Current Medicinal Chemistry. 2015;**22**(14):1757-1773

[36] Dufour N, Rao RP. Secondary metabolites and other small molecules as intercellular pathogenic signals. FEMS Microbiology Letters. 2011;**314**(1):10-17

[37] Barraud N, Hassett DJ, Hwang SH, Rice SA, Kjelleberg S, Webb JS. Involvement of nitric oxide in biofilm dispersal of Pseudomonas aeruginosa. Journal of Bacteriology. 2006;**188**(21):7344-7353

[38] Wolfmeier H, Pletzer D, Mansour SC, Hancock REW. New perspectives in biofilm eradication. ACS Infectious Diseases. 2018;**4**(2):93-106

[39] Lansdown ABG. Silver in health care: Antimicrobial effects and safety in use. Current Problems in Dermatology. 2006;**33**:17-34

[40] Lemire JA, Kalan L, Gugala N, Bradu A, Turner RJ. Silver oxynitrate-an efficacious compound for the prevention and eradication of dual-species biofilms. Biofouling. 2017;**33**(6):460-469

[41] Okshevsky M, Regina VR, Meyer RL. Extracellular DNA as a target for biofilm control. Current Opinion in Biotechnology. 2015;**33**:73-80

[42] Tetz GV, Artemenko NK, Tetz VV. Effect of DNase and antibiotics on biofilm characteristics. Antimicrobial Agents and Chemotherapy. 2009;**53**(3):1204-1209

[43] Whitchurch CB, Tolker-Nielsen T, Ragas PC, Mattick JS. Extracellular DNA

required for bacterial biofilm formation. Science. 2002;**295**(5559):1487

[44] Sultana ST, Babauta JT, Beyenal H. Electrochemical biofilm control: A review. Biofouling. 2015;**31**(9-10):745-758

[45] Freebairn D, Linton D, Harkin-Jones E, Jones DS, Gilmore BF, Gorman SP. Electrical methods of controlling bacterial adhesion and biofilm on device surfaces. Expert Review of Medical Devices. 2013;**10**(1):85-103

[46] Sabelnikov AG, Cymbalyuk ES, Gongadze G, Borovyagin VL. Escherichia coli membranes during electrotransformation: An electron microscopy study. Biochimica et Biophysica Acta. 1991;**1066**(1):21-28

[47] Hancock REW, Rozek A. Role of membranes in the activities of antimicrobial cationic peptides. FEMS Microbiology Letters. 2002;**206**(2):143-149

[48] Sultana ST, Atci E, Babauta JT, Mohamed Falghoush A, Snekvik KR, Call DR, et al. Electrochemical scaffold generates localized, low concentration of hydrogen peroxide that inhibits bacterial pathogens and biofilms. Scientific Reports. 2015;**5**(1):14908

[49] Chen L, Wen Y. The role of bacterial biofilm in persistent infections and control strategies. International Journal of Oral Science. 2011;**3**(2):66-73

[50] Renner LD, Weibel DB. Physicochemical regulation of biofilm formation. MRS Bulletin. 2011;**36**(5):347-355

[51] Kalia VC, Patel SKS, Kang YC, Lee JK. Quorum sensing inhibitors as antipathogens: Biotechnological applications. Biotechnology Advances. 2019;**37**(1):68-90

[52] Otani S, Hiramatsu K, Hashinaga K, Komiya K, Umeki K, Kishi K, et al. Sub-minimum inhibitory concentrations of ceftazidime inhibit Pseudomonas aeruginosa biofilm formation. Journal of Infection and Chemotherapy. 2018;**24**(6):428-433

[53] Klinger-Strobel M, Stein C, Forstner C, Makarewicz O, Pletz MW. Effects of colistin on biofilm matrices of Escherichia coli and Staphylococcus aureus. International Journal of Antimicrobial Agents. 2017;**49**(4):472-479

[54] Butini ME, Cabric S, Trampuz A, Di Luca M. In vitro anti-biofilm activity of a biphasic gentamicin-loaded calcium sulfate/hydroxyapatite bone graft substitute. Colloids and Surfaces B: Biointerfaces. 2018;**161**:252-260

Section 2

Bacterial Pathogenesis and Biofilms

Chapter 4

Campylobacter: Virulence Factors and Pathogenesis

Matthew Terzungwe Tion, Kenneth Ikejiofor Ogbu and Felix Kundu Shima

Abstract

The species from the genus Campylobacter are the common causes of foodborne bacterial pathogens found worldwide. The diseases that arise from the infection of this bacterial agent are sometimes self-limiting or can range from mild symptoms to fatal illnesses. The disease is reported in more than 500 million cases of diarrhea annually. The taxonomy, pathogenesis and treatment of Campylobacter is been discussed here. Several virulence factors of Campylobacter are involved in playing a crucial role in pathogenesis, e.g., the chemotactically controlled cellular motility, the bacterial adhesion, the invasion into the host cell, and toxin formation. When a specific diagnosis is made, antibiotic therapy is advocated for use to reduce symptoms. The random use of antibiotics in the treatment of infectious diseases has brought about the emergence of many antibiotic-resistant bacteria, which have become a public health problem and a menace to society.

Keywords: adhesion, bacteria, Campylobacter, gastroenteritis, virulence factors

1. Introduction

Campylobacter was first described in 1913 [1] but was initially classified in the genus *Vibrio*. With its similarities to the Helicobacter genus, it was finally grouped into the genus Campylobacter (family Campylobacteraceae, order Campylobacterales, class Epsilonproteobacteria, phylum Proteobacteria), which consisted of 33 species and subspecies, showing a broad ecological distribution [2].

Campylobacter is one of the most common causes of bacterial gastroenteritis worldwide [3]. Currently, there are several species within the genus Campylobacter divided into 43 child taxa with a validly published species (http://www.bacterio.net/campylobacter.html) [4] as shown in **Table 1**. The most common sources of transmission of its infection are contaminated water, raw or contaminated milk, and food especially undercooked meat or meat products [5].

The taxonomic classification of Campylobacter species has been long characterized by the phenotype of bacterial isolates, where molecular characterization played a minute role in the description of bacteria [6]. There is a drastic change involving all the branches of taxonomy with the advent of DNA sequencing technologies, the

S/No	Campylobacter specie
1	*Campylobacter sputorum*
2	*Campylobacter jejuni*
3	*Campylobacter fetus*
4	*Campylobacter coli*
5	*Campylobacter concisus*
6	*Campylobacter laridis*
7	*Campylobacter mucosalis*
8	*Campylobacter hyointestinalis*
9	*Campylobacter curvus*
10	*Campylobacter upsaliensis*
11	*Campylobacter rectus*
12	*Campylobacter helveticus*
13	*Campylobacter showae*
14	*Campylobacter gracilis*
15	*Campylobacter lanienae*
16	*Campylobacter hominis*
17	*Campylobacter insulaenigrae*
18	*Campylobacter canadensis*
19	*Campylobacter peloridis*
20	*Campylobacter cuniculorum*
21	*Campylobacter avium*
22	*Campylobacter ureolyticus*
23	*Campylobacter subantarcticus*
24	*Campylobacter volucris*
25	*Campylobacter corcagiensis*
26	*Campylobacter iguaniorum*
27	*Campylobacter hepaticus*
28	*Campylobacter geochelonis*
29	*Campylobacter pinnipediorum*
30	*Campylobacter ornithocola*
31	*Campylobacter blaseri*
32	*Campylobacter armoricus*
33	*Campylobacter novaezeelandiae*
34	*Campylobacter vulpis*
35	*Campylobacter taeniopygiae*
36	*Campylobacter portucalensis*
37	*Campylobacter massiliensis*
38	*Campylobacter estrildidarum*

S/No	Campylobacter specie
39	*Campylobacter aviculae*
40	*Campylobacter bilis*
41	*Campylobacter anatolicus*
42	*Campylobacter suis*
42	*Campylobacter majalis*
http://www.bacterio.net/campylobacter.html.	

Table 1.
Currently described Campylobacter species.

standards for the description of Campylobacter species have been updated to a biphasic approach with genotype and phenotype descriptions both being important [7].

DNA-based classifications also have no universal standard and different methodologies (eg. multi-locus sequence typing vs. whole genome sequencing) that allow different resolving capacities to distinguish between strain variants as well as dealing with complex evolutionary phenomena such as recombination in different ways. Nonetheless, taxonomic updates based on DNA sequences are essential and have led to the inclusion of species previously not classified as Campylobacter [8] and the exclusion and reclassification of others [9].

The members of the Campylobacter genus share so many common features. Morphologically, Campylobacter is a slightly curved or spiral, rod-shaped bacteria with single, bipolar, or entire absence flagellum depending on the specie, motile and non-spore-forming, obligate microaerophilic (requires minimum 5% O2 level, Nitrogen 85%, 10% CO2) heat labile, thermophilic, Gram-negative bacteria and exhibits optimum growth at 42° C [10–13]. An atmosphere containing increased hydrogen is required by some species for microaerobic growth [10]. Campylobacter species measure approximately 0.2 to 0.8 by 0.5 to 5 m in size and are chemoorganotrophs which acquire their energy sources from amino acids or tricarboxylic acid cycle intermediates [14].

This bacterium takes residence in different places, commonly the gastrointestinal tracts of many animal species where it serves as commensal or pathogenic [15]. Most infections in humans are attributed to *Campylobacter jejuni* and *Campylobacter coli*, additional species are isolated from humans too [13]. *C. jejuni* is the major causative agent of human foodborne gastroenteritis globally resulting in more than 500 million cases of diarrhea annually [16, 17]. In severe cases of infection incriminated by *C. jejuni*, there is a development of post-infection complications such as Guillain Barré Syndrome [18].

The genome size of *C. jejuni* and *C. coli* is approximately 1600 ± 1700 kilobases (kb). This is comparatively small in relation to the enteropathogens such as *Escherichia coli* with a genome size of approximately 4500 kb [19, 20]. Since the genome of *C. jejuni* was sequenced, it revealed the presence of hypervariable sequences that consist of homopolymeric tracts [21].

Among the species of Campylobacter, *C. jejuni* show a high level of variability within many sequences, helping it to adapt to different harsh environments [22]. Within these hypervariable genes, several phenotypic variations and vast diversity within Campylobacter populations occur, especially after transmission from animals or humans. This is shown in different experiments performed in vivo and in vitro, resulting in several genetic changes and mutations that contribute to a high genetic diversity [23, 24]. When samples collected from animals and humans are sequenced and typed, only a few genotypes are similar bringing about strain-dependent pathogenicity and specific colonization ability [25, 26].

2. Pathogenesis

Campylobacter is very infectious as low infective doses of 500 to 800 CFU can cause a problem in humans [27, 28]. Thermotolerant campylobacters, such as *C. Jejuni* and *C. coli*, are the most frequent cause of bacterial infection of the lower intestine worldwide [29]. The mechanism of pathogenesis of *C. jejuni* comprises of four main stages: adhesion to intestinal cells, colonization of the digestive tract, invasion of targeted cells, and toxin production [30].

Several virulence factors of Campylobacter are involved in playing a crucial role in pathogenesis, e.g., the chemotactically controlled cellular motility, the bacterial adhesion, the invasion into the host cell, and toxin formation. In addition to the roles of virulence factors in host colonization, additional factors are involved in successful colonization, such as various genes, antigens, mechanisms of iron utilization, and the response to oxidative and environmental stress. The poor knowledge in understanding, which bacterial and cellular factors are, involved in pathogenicity is not only due to the genetic inter- and intrastrain variability but also to differences between the laboratory strains and the different host cell lines and protocols used in the different laboratories [31]. However, even if the exact mechanism of infection in humans is not yet known, three basic steps can be identified [32].

First, the colonization of the intestinum, especially the crypts of the gut mucosa, a specific adhesion occurs to proteins of the host epithelium, followed by the invasion of the intestinal cells and the translocation of the bacterium, either trans- or paracellularly. At this point, Campylobacter multiplies in the intestinal mucosa releasing toxins that necrotize the intestinal villi. The damage to the intestinal epithelium results in a loss of function, opening of the shielding barrier and the tight junctions, induction of inflammation, release of electrolytes from the systemic compartment of the host to the lumen of the gut, and finally to strong and bloody diarrhea. Furthermore, the adhesion of the bacteria to the epithelial cells is accompanied by a strong pro-inflammatory immune response [33].

2.1 The mechanism of pathogenesis

Humans get infected primarily by contact with live animals or through consumption of contaminated foodstuffs, contact with live poultry, consumption of poultry meat [34, 35], Pork meat [36], beef [37], drinking water from untreated water sources, and raw milk [38].

For adequate attachment to a host, microorganisms require adherence factors which are usually surface appendages such as the pili that are located on the surface of many Gram-negative and Gram-positive species. Genome annotations of several *C. jejuni* strains do not include obvious pilus or pilus-like open reading frames [21, 38]. A multi-protein type II-like secretion system of a type that is associated with pilus assembly in *Vibrio cholerae* and *Neisseria gonorrhoeae* was identified as part of the competence machinery, but an actual pilus-like structure has not been identified [39].

Campylobacter colonization of the host mechanism involves primary intestinal cells – Islets of cobblestone cells. The intestinal mucus attenuates *C. jejuni* invasion in-vitro in chicken. A specific avian intestinal factor rather than tissue tropism underlies Campylobacter commensalism in Chickens [40]. Upon infection, Campylobacter elicits the secretion of Interlukin-8, IL-8 and Cytokine. The bacteria adhere preferentially to mucus overlying the intestinal tissue. There is also an interaction of *C. jejuni* with tissue or mucus via the flagella [41]. Several factors have been identified as influencing the binding of Campylobacter to epithelial cells of the host in-vitro [30].

Other proteins involved in campylobacter virulence and adhesions to the host call are the CadF (Campylobacter adhesin to fibronectin) and Peb1 proteins. Inactivation of the CadF gene seemed to render *C. jejuni* capable of colonizing the cecum of chicks. Also, CadF protein is required for optimal bacterial adhesion to the extracellular matrix and the colonization of newly hatched Leghorn chicken [42].

CadF binds specifically to fibronectin, which is located basolaterally on epithelial cells in situ [43, 44]. CadF is required for maximal binding and invasion by *C. jejuni* in vitro, and cadF mutants are greatly reduced in chick colonization compared with the wild type [44].

Another characterized adhesion is the JlpA, a surface exposed lipoprotein that is highly required for HEp-2 cell binding [45]. JlpA binds to Hsp90α, some of which is surface localized in these cells [46]. The process of JlpA binding to Hsp90α activates NF-κB and p38 mitogen-activated protein (MAP) kinase, both of them contribute to proinflammatory responses [46]. This is an indication that some of the inflammatory processes that are observed during the pathogenesis of *C. jejuni* might be related to JlpA-dependent adherence. Another lipoprotein, CapA, was implicated as a possible adhesion. CapA is an autotransporter that is homologous to an autotransporter adhesin, and CapA-deficient mutants have decreased adherence to Caco-2 cells and decreased colonization and persistence in a chick model [47].

Some putative adhesins of *C. jejuni* are located in the periplasm such as the Peb1 adhesin, also known as CBF1 required for adherence to HeLa cells [48, 49]. Peb1 shares homology with the periplasmic-binding proteins of amino acid ATP-binding cassette (ABC) transporters [50, 51]. Peb1 binds to both aspartate and glutamate with high affinity, and peb1-deficient mutants cannot grow if these amino acids are the major carbon source [51]. Although Peb1 has not been localized to the inner or outer membrane, some has been observed in culture supernatants [51]. Furthermore, Peb1 contains a predicted signal peptidase II recognition site, a common motif in surface-localized lipoproteins, and so there is a possibility that some Peb1 is surface accessible, despite the failure of fractionation techniques to demonstrate this [50, 51]. Mutants that lack peb1 colonize mice poorly, but this could be attributed to the loss of either the adhesion or the amino-acid-transport functions, or both [49, 51].

The cytotoxic activity in *C. jejuni* is associated with a cytolethal-distending toxin (CDT) [52]. CDT induces DNA double-strand breaks leading to cell cycle arrest in the G2 phase and provokes cell distension and eventually cell death [52–54]. CDT also seems to play a role in the invasion and/or survival of *C. Jejuni* in HeLa cells [55].

The growth temperature of the bacteria significantly affects the ability of *C. jejuni* to bind to epithelial cell lines in-vitro. Its maximum adhesion to INT-407 is 37°C [56]. Generally, the binding of *C. jejuni* to cultured cells is not affected by temperature or the Phylogenetic origin of the target cell. The number of bacteria in the inoculum or multiplicity of infection regulates the ability of Campylobacter to invade [57].

The Cytotoxic effect is characterized by remarkable cell distension that is obvious 48 to 72 h after the addition of bacteria-free supernatant and results in cell death. This cell distension is evident in the appearance of HeLa cells, which are star-shaped [53].

Adhesion, invasion, and cytotoxic assay indicate that the ability to invade and induce IL-8 production, to produce CDT, and to resist bile salt is widespread among *C. jejuni* isolates [58], nevertheless, a higher degree of bile salt resistance and more. Pronounced CDT productions are associated with strains causing enteritis in humans. Furthermore, the CheY appears to be a modulator of *C. jejuni* virulence.

Molecular studies revealed a high rate of variation of homopolymeric runs – lipooligosaccharides, capsules, or flagellin that are responsible for virulence, are also important for the survival strategy of *C. jejuni* [21]. The flagellar apparatus is more important for the invasion and translocation of *C. jejuni* in contact with the host cell and for Chicken gut colonization. The Lipooligosaccharide (LOS) of *C. jejuni* is highly variable but their structures resemble human neuronal gangliosides. It is thought that this phenomenon results in autoimmune disorders, including Guillain–Barré syndrome (GBS), a paralytic neuropathy that occurs following approximately 1 in every 1000 cases of campylobacterosis, and Miller–Fisher syndrome, a variant of GBS. Much research works have been done to advance the understanding of the mechanism by which *C. Jejuni* infections result in the conditions above [59–62].

The presence of variation in the capsule structure observed has been linked to multiple mechanisms that include phase variation of structural genes and an O-methyl phosphoramidate modification [63–66]. Many strains of *C. jejuni* are thought to produce both LOS and a high molecular weight lipopolysaccharide (HMW LPS) that is a highly variable capsular polysaccharide. The structural capsules of several *C. jejuni* strains have been determined including strain 11,168 and strain RM1221 showing their similarities and differences in structure. The former presented with 6-methyl-d-glycero-α-l-glucoheptose, β-d-glucouronic acid modified with 2-amino-2-deoxyglycerol, β-d-GalfNAc and β-d-ribose [64], and contains a novel modification on the GalfNAc [65], the latter having 6-deoxy-d-manno-heptose and d-xylose [67], which are two sugars that are not often detected in bacterial polysaccharides while some strains possess teichoic acid-like or hyaluronic acid-like capsules [68, 69]. The *C. jejuni* capsule is responsible for serum resistance, the adherence and invasion of epithelial cells, chick colonization and virulence in a ferret model [70–72].

Other important genes -CiaB (Campylobacter invasion antigen B) is involved in mutagenesis and is required for the secretion process and effective entry of the bacterium into the host cell [73–75]; while the Cjl121c is essential for host colonization and virulence [76].

Mutation experiments showed that many genes involved in Campylobacter virulence are conserved across species (e.g., CadF, Peb1, jlpA, cdt operon, CiaB, and flagellin genes). Among the same strains of the same origin, there are differences in virulence characteristics [77].

Campylobacteriosis in humans presents symptoms including diarrhea (often bloody), cramping, abdominal pain, nausea, and headaches [78] most commonly caused by the species *C. jejuni* and *C. coli* [79, 80]. It has also been tagged as an 'emerging *Campylobacter spp*.' [13], including *Campylobacter concisus* [81]. *Campylobacter sputorum* [82], *Campylobacter upsaliensis* [82], *Campylobacter ureolyticus* [83] and *Campylobacter hyointestinalis* [84]. *C. jejuni*, *C. coli*, *Campylobacter lari*, and *C. upsaliensis* form a genetically close group referred to as the thermotolerant campylobacters because they grow optimally at 42°C [85] and the remaining Campylobacter species are classified into other general groups [86]. The increasing availability of genetic and genomic data on the species' characteristics and ecological associations due to the improved diagnostic technologies has transformed our perception of the clinical importance of "emerging *Campylobacter spp*." [84].

Generally, Campylobacter causes a self-limiting clinical illness that lasts 5 to 7 days; the infection resolves without the use of antimicrobials in the majority of cases but 5–10% of patients relapse after their initial illness [87]. However, the infection can take a much more severe course, especially in infants, elderly people, and immunosuppressed patients (e.g., HIV), so intensified antibiotic treatment is necessary [88].

3. Treatment

The culture-independent diagnostic tests (CIDT) detect the presence of specific antigens or DNA sequences and are recently more used for the detection of bacterial enteric infections such as Campylobacter, Salmonella, Shigella, Shigatoxin–producing *E. coli*, Vibrio, and Yersinia [89].

Various PCR methods are used for the specific identification of Campylobacter e.g. single locus sequencing of the flagellar flaA and flaB genes and Multilocus sequence typing. Campylobacter-specific genome sections can also be detected by multiplex PCR [90].

Standardized molecular typing methods such as pulsed-field gel electrophoresis [91] and flagellin typing (fla typing) by restriction fragment length polymorphism analysis of a PCR product [92] are in use worldwide.

The random use of antibiotics in the treatment of infectious diseases has brought about the emergence of many antibiotic-resistant bacteria, which have become a public health problem and a menace to society. In recent years, several studies have reported the problem of antibiotic resistance in the various strains of bacteria [93, 94].

In self-limiting infection, no treatment is required but uncomplicated enterocolitis is managed symptomatically with fluid therapy consisting of electrolytes and volume substitution [95–97].

In cases where a specific diagnosis is made and a severe progression to a fatal illness, immunosuppression, or lack of improvement of symptoms antibiotic therapy with macrolides (azithromycin), fluoroquinolones (ciprofloxacin), and tetracyclines is recommended. Resistance testing should be performed routinely for these cases. The use of cephalosporins should be highly avoided due to high resistance rates. Macrolides, fluoroquinolones and aminoglycosides are classified as critically important antimicrobials, while tetracycline is considered a highly important antimicrobial [98]. Resistance to (fluoro)quinolones and tetracyclines is highly prevalent in *Campylobacter spp.* isolates, while resistance to erythromycin is typically low to moderate [99–101].

Macrolides are the first-line antibiotic for the treatment of enteric gastroenteritis, while fluoroquinolones and tetracyclines remain as alternatives [102–104]. Systemic infections are routinely treated with aminoglycosides [104, 105] with low resistance [100].

Fluoroquinolones act by primarily targeting the DNA gyrase [106]. DNA gyrase is a heterotetrameric type IIA topoisomerase, consisting of two polypeptide subunits (GyrA and GyrB, encoded by gyrA and gyrB, respectively), catalyzing ATP-dependent negative supercoiling of DNA to regulate replication, repair and gene expression [107–109]. Resistance to (fluoro)quinolones among *Campylobacter spp.* is largely mediated by chromosomal mutations in the quinolone resistance-determining region (QRDR) of gyrA, typically conferred by the C257T nucleotide mutation (Thr-86-Ile) [110].

The prevalence of Campylobacter gastroenteritis can be significantly reduced by risk-based vaccination, although, there is no commercial vaccine available at the moment, which is also due to the great antigenic diversity of the bacterium. A capsule polysaccharide-based vaccine has proven successful against diarrhea in primates [111]. In order to increase its immunogenicity in humans, it has been coupled to liposomes [112].

4. Conclusion

It is established that Infections caused by Campylobacter are one of the major food-borne causes of gastroenteritis, which can range from mild symptoms to fatal illness worldwide. With the advancement in molecular technology, the epidemiology of some Campylobacter infections remains a mystery.

Care must be taken in severe forms, especially in infants, elderly people, and immunosuppressed patients (e.g., HIV), and must be placed on antibiotic therapy bearing in mind that random use of antibiotics has brought about antibiotic-resistant strains of the bacteria. Further research work is needed to unveil more information.

Conflict of interest

The authors declare no conflict of interest.

Author details

Matthew Terzungwe Tion[1*], Kenneth Ikejiofor Ogbu[2] and Felix Kundu Shima[3]

1 University of Agriculture, Makurdi, Benue State, Nigeria

2 Federal College of Animal Health and Production Technology, National Veterinary Research Institute, Vom, Plateau State, Nigeria

3 Ministry of Agriculture and Natural Resources, Makurdi, Benue State, Nigeria

*Address all correspondence to: tions_doc@yahoo.co.uk

References

[1] McFadyean J, Stockman S. Report of the Departmental Committee Appointed by the Board of Agriculture and Fisheries to Inquire into Epizootic Abortion. Part III. HMSO London: Abortion in Sheep; 1913

[2] Parte AC List of prokaryotic names with standing in nomenclature. 2018. https://www.bacterio.net/. [Accessed: August 18, 2020]

[3] Kaakoush NO, Castano-Rodriguez N, Mitchell HM, Man SM. Global epidemiology of Campylobacter infection. Clinical Microbiology Reviews. 2015;**28**:687-720. DOI: 10.1128/CMR.00006-15

[4] Aydin F, Abay S, Kayman T, Karakaya E, Mustak HK, Mustak IB, et al. *Campylobacter anatolicus* sp. nov., a novel member of the genus *Campylobacter* isolated from feces of Anatolian ground squirrel (Spermophilus xanthoprymnus) in Turkey. Systematic and Applied Microbiology. 2021;**44**:126265

[5] World Health Organization (WHO). Camplyobacter. 2020. https://www.who.int/news-room/fact-sheets/detail/campylobacter

[6] Ursing JB, Lior H, Owen RJ. Proposal of minimal standards for describing new species of the family Campylobacteraceae. International Journal of Systematic Bacteriology. 1994;**44**:842-845. DOI: 10.1099/00207713-44-4-842

[7] On SLW, Miller WG, Houf K, Fox JG, Vandamme P. Minimal standards for describing new species belonging to the families Campylobacteraceae and Helicobacteraceae: Campylobacter, Arcobacter, helicobacter and Wolinella spp. International Journal of Systematic and Evolutionary Microbiology. 2017;**67**(12):5296-5311. DOI: 10.1099/ijsem.0.002255

[8] Vandamme P, Debruyne L, De Brandt E, Falsen E. Reclassifcation of Bacteroides ureolyticus as Campylobacter ureolyticus comb. nov., and emended description of the genus Campylobacter. International Journal of Systematic and Evolutionary Microbiology. 2010;**2010**(60):2016-2022. DOI: 10.1099/ijs.0.017152-0

[9] Goodwin CS et al. Transfer of Campylobacter pylori and Campylobacter mustelae to helicobacter gen. Nov. as helicobacter pylori comb. nov. and helicobacter mustelae comb. nov., respectively. International Journal of Systematic and Evolutionary Microbiology. 1989;**1989**(39):397-405. DOI: 10.1099/00207713-39-4-397

[10] Vandamme P, Falsen E, Rossau R, Hoste B, Segers P, Tytgat R, et al. Revision of Campylobacter, helicobacter, and Wolinella taxonomy: Emendation of generic descriptions and proposal of Arcobacter gen. Nov. International Journal of Systematic Bacteriology. 1991b;**41**:88-103

[11] Newell DG, Fearnley C. Sources of Campylobacter colonization in broiler chickens. Applied and Environmental Microbiology. 2003;**69**(8):4343-4351

[12] Koenraad PM, Jacobs-Reitsma WF, van der Laan T, Beumer RR, Rombouts FM. Antibiotic susceptibility of Campylobacter isolates from sewage and poultry abattoir drain water. Epidemiology and Infection. 1995;**115**:475-483

[13] Man SM. The clinical importance of emerging Campylobacter species. Nature

Reviews Gastroenterology & Hepatology. 2011;**8**(12):669-685

[14] Vandamme P, Dewhirst FE, Paster BJ, On SLW, Campylobacteraceae. . In: Garrity GM, Brenner DJ, Krieg NR, Staley JT, editors. Bergey's Manual of Systematic Bacteriology. Vol. 2. New York, NY: Springer Science; 2005. pp. 1147-1160

[15] Iraola G et al. Genomic evidence for the emergence and evolution of pathogenicity and niche preferences in the genus Campylobacter. Genome Biology and Evolution. 2014;**4**:2392-2405. DOI: 10.1093/gbe/evu195

[16] Centers for Disease Control and Prevention-CDC. National Center for emerging and zoonotic Infectius disease. Campylobacteriosis. 2020;**1**:720. https://wwwnc.cdc.gov/travel/yellowbook/2020/travel-related-infectious-diseases/campylobacteriosis. [Accessed: January 21, 2021]

[17] European Food Safety Authority-EFSA. The European Union summary report on antimicrobial resistance in zoonotic and indicator bacteria from humans, animals and food in 2017/2018. EFSA J. Parma Italy. 2020;**18**(3):6007. DOI: 10.2903/j.efsa.2020.6007

[18] Goodfellow J, Willison H. Guillain–Barré syndrome: A century of progress. Nature Reviews. Neurology. 2016;**12**:723-731. DOI: 10.1038/nrneurol.2016.172

[19] Chang N, Taylor DE. Use of pulsed-field agarose gel electrophoresis to size genomes of Campylobacter species and to construct a SalI map of Campylobacter jejuni UA580. Journal of Bacteriology. 1990;**172**:5211-5217

[20] Nuijten PJM, Bartels C, Bleumink-Pluym NMC, Gaastra W, van der Zeijst BAM. Size and physical map of the Campylobacter jejuni chromosome. Nucleic Acids Research. 1990a;**18**:6211-6214

[21] Parkhill J, Wren BW, Mungall K, Ketley JM, Churcher C, Basham D, et al. The genome sequence of the food-borne pathogen Campylobacter jejuni reveals hypervariable sequences. Nature. 2000;**403**:665-668

[22] Dasti JI, Tareen AM, Lugert R, Zautner AE, Gross U. Campylobacter jejuni: A brief overview on pathogenicityassociated factors and disease-mediating mechanisms. International Journal of Medical Microbiology. 2010;**300**:205-211

[23] Abd El-Hamid MI, Abd El-Aziz NK, Samir M, El-Naenaeey E-SY, Abo Remela EM, Mosbah RA, et al. Genetic diversity of Campylobacter jejuni isolated from avian and human sources in Egypt. Frontiers in Microbiology. 2019;**10**:1-14

[24] Mohawk KL, Poly F, Sahl JW, Rasko DA, Guerry P. High frequency, spontaneous motA mutations in Campylobacter jejuni strain 81-176. PLoS One. 2014;**9**:1-8

[25] deBoer P, Wagenaar JA, Achterberg R, van Putten J, Schouls L, Duim B. Generation of Campylobacter jejuni genetic diversity in vivo. Molecular Microbiology. 2002;**44**:351-359

[26] Ridley AM, Toszeghy MJ, Cawthraw SA, Wassenaar TM. Newell DG genetic instability is associated with changes in the colonization potential of Campylobacter jejuni in the avian intestine. Journal of Applied Microbiology. 2008;**105**:95-104

[27] Janssen R, Krogfelt KA, Cawthraw SA, van Pelt W, Wagenaar JA, Owen RJ. Host-pathogen interactions

in Campylobacter infections: The host perspective. Clinical Microbiology Reviews. 2008;**21**:505-518

[28] Robinson DA. Infective dose of Campylobacter jejuni in milk. British Medical Journal (Clinical Research Ed.). 1981;**282**:1584

[29] Miller WG, Mandrell RE. Prevalence of Campylobacter in the food and water supply: Incidence, outbreaks, isolation and detection. In: Konkel ME, Ketley JM, editors. Campylobacter, Molecular and Cellular Biology. Wymondham, UK: Horizon Bioscience; 2005. pp. 101-163

[30] Haddad N, Marce C, Magras C, Cappelier J-M. An overview of methods used to clarify pathogenesis mechanisms of Campylobacter jejuni. Journal of Food Protection. 2010;**73**(4):786-802

[31] Poli VFS, Thorsen L, Olesen I, Wik MT, Jespersen L. Differentiation of the virulence potential of Campylobacter jejuni strains by use of gene transcription analysis and a Caco-2 assay. International Journal of Food Microbiology. 2012;**155**:60-68

[32] Konkel ME, Monteville MR, Rivera-Amill V, Jones LA. The pathogenesis of-Campylobacter jejuni-mediated enteritis. Current Issuse in Interest Mocrobiology. 2001;**2**:55-71

[33] Aguilar C, Jiménez-Marín Á, Martins RP, Garrido JJ. Interaction between Campylobacter and intestinal epithelial cells leads to a different proinflammatory response in human and porcine host. Veterinary Immunology and Immunopathology. 2014;**162**:14-23

[34] Hanninen ML, Perko-Makela P, Pitkala A, Rautelin H. A three-year study of Campylobacter jejuni genotypes in humans with domestically acquired infections and in chicken samples from the Helsinki area. Journal of Clinical Microbiology. 2000;**38**:1998-2000

[35] Pearson AD, Greenwood MH, Donaldson J, Healing TD, Jones DM, Shahamat M, et al. Continuous source outbreak of campylobacteriosis traced to chicken. Journal of Food Protection. 2000;**63**:309-314

[36] Fosse J, Seegers H, Magras C. Foodborne zoonoses due to meat: A quantitative approach for a comparative risk assessment applied to pig slaughtering in Europe. Veterinary Research. 2008;**39**:1-16

[37] Fosse J, Seegers H, Magras C. Hie'rarchiser les risques de zoonoses alimentaires: une approche quantitative. Application aux dangers bacte'riens transmis par les viandes porcine et bovine. Revue scientifique et technique Official International Epizootic. 2008;**27**:643-655

[38] Engberg J, Gerner-Smidt P, Scheutz F, On SL, Moller Nielsen E, Molbak K. Water-borne Campylobacter jejuni infection in a Danish town—A 6-week continuous source outbreak. Clinical Microbiology and Infection. 1998;**4**:648-656

[39] Wiesner RS, Hendrixson DR, DiRita VJ. Natural transformation of Campylobacter jejuni requires components of a type II secretion system. Journal of Bacteriology. 2003;**185**:5408-5418

[40] Byrne CM, Clyne M, Bourke B. Campylobacter jejuni adhere to and invade chicken intestinal epithelial cells in vitro. Microbiology. 2007;**153**:561-569

[41] MacCallum AJ, Harris D, Haddock G, Everest PH. Campylobacter jejuni–infected human epithelial cell lines vary in their ability to secrete interleukin-8

compared to in vitro–infected primary human intestinal tissue. Microbiology. 2006;**152**:3661-3665

[42] Ziprin RL, Young CR, Stanker LH, Hume ME, Konkel ME. The absence of cecal colonization of chicks by a mutant of Campylobacter jejuni not expressing bacterial fibronectinbinding protein. Avian Diseases. 1999;**43**:586-589

[43] Monteville MR, Konkel ME. Fibronectinfacilitated invasion of T84 eukaryotic cells by Campylobacter jejuni occurs preferentially at the basolateral cell surface. Infection and Immunity. 2002;**70**:6665-6671

[44] Monteville MR, Yoon JE, Konkel ME. Maximal adherence and invasion of INT 407 cells by Campylobacter jejuni requires the CadF outermembrane protein and microfilament reorganization. Microbiology. 2003;**149**:153-165

[45] Jin S et al. JlpA, a novel surface-exposed lipoprotein specific to Campylobacter jejuni, mediates adherence to host epithelial cells. Molecular Microbiology. 2001;**39**:1225-1236

[46] Jin S, Song YC, Emili A, Sherman PM, Chan VL. JlpA of Campylobacter jejuni interacts with surface-exposed heat shock protein 90α and triggers signalling pathways leading to the activation of NF-κB and p38 MAP kinase in epithelial cells. Cellular Microbiology. 2003;**5**:165-174

[47] Ashgar SS et al. CapA, an autotransporter protein of Campylobacter jejuni, mediates association with human epithelial cells and colonization of the chicken gut. Journal of Bacteriology. 2007;**189**:1856-1865

[48] Kervella M et al. Isolation and characterization of two Campylobacter glycine-extracted proteins that bind to HeLa cell membranes. Infection and Immunity. 1993;**61**:3440-3448

[49] Pei Z et al. Mutation in the peb1A locus of Campylobacter jejuni reduces interactions with epithelial cells and intestinal colonization of mice. Infection and Immunity. 1998;**66**:938-943

[50] Pei Z, Blaser MJ. PEB1, the major cell-binding factor of Campylobacter jejuni, is a homolog of the binding component in gram-negative nutrient transport systems. The Journal of Biological Chemistry. 1993;**268**:18717-18725

[51] Leon-Kempis Mdel R, Guccione E, Mulholland F, Williamson MP, Kelly DJ. The Campylobacter jejuni PEB1a adhesin is an aspartate/glutamatebinding protein of an ABC transporter essential for microaerobic growth on dicarboxylic amino acids. Molecular Microbiology. 2006;**60**:1262-1275

[52] Purdy D, Buswell CM, Hodgson AE, McAlpine K, Henderson I, Leach SA. Characterisation of cytolethal distending toxin (CDT) mutants of Campylobacter jejuni. Journal of Medical Microbiology. 2000;**49**:473-479

[53] Cortes-Bratti X, Frisan T, Thelestam M. The cytolethal distending toxins induce DNA damage and cell cycle arrest. Toxicon. 2001;**39**:1729-1736

[54] Thelestam M, Frisan T. Cytolethal distending toxins. In: Aktories K, Just I, editors. Special Issue on Emerging Bacterial Toxins. Vol. 152. Berlin: Springer; 2005. pp. 111-133

[55] Biswas D, Fernando U, Reiman C, Willson P, Potter A, Allan B. Effect of cytolethal distending toxin of

Campylobacter jejuni on adhesion and internalization in cultured cells and in colonization of the chicken gut. Avian Diseases. 2006;**50**:586-593

[56] Konkel ME, Corwin MD, Joens LA, Cieplak W. Factors that influence the interaction of Campylobacter jejuni with cultured mammalian cells. Journal of Medical Microbiology. 1992;**37**:30-37

[57] Hu L, Kopecko DJ. Campylobacter jejuni 81-176 associates with microtubules and dynein during invasion of human intestinal cells. Infection and Immunity. 1999;**67**:4171-4182

[58] Van Deun K, Haesebrouck F, Heyndrickx M, Favoreel H, Dewulf J, Ceelen L, et al. Virulence properties of Campylobacter jejuni isolates of poultry and human origin. Journal of Medical Microbiology. 2007;**56**:1284-1289

[59] Nachamkin I. Chronic effects of Campylobacter infection. Microbes and Infection. 2002;**4**:399-403

[60] Hughes R. Campylobacter jejuni in Guillain–Barre syndrome. Lancet Neurology. 2004;**3**:644

[61] Komagamine T, Yuki N. Ganglioside mimicry as a cause of Guillain–Barre syndrome. CNS & Neurological Disorders Drug Targets. 2006;**5**:391-400

[62] Yu RK, Usuki S, Ariga T. Ganglioside molecular mimicry and its pathological roles in Guillain–Barre syndrome and related diseases. Infection and Immunity. 2006;**74**:6517-6527

[63] Karlyshev AV, Linton D, Gregson NA, Lastovica AJ, Wren BW. Genetic and biochemical evidence of a Campylobacter jejuni capsular polysaccharide that accounts for Penner serotype specificity. Molecular Microbiology. 2000;**35**:529-541

[64] St MF et al. The structures of the lipooligosaccharide and capsule polysaccharide of Campylobacter jejuni genome sequenced strain NCTC 11168. European Journal of Biochemistry. 2002;**269**:5119-5136

[65] Szymanski CM et al. Detection of conserved N-linked glycans and phase-variable lipooligosaccharides and capsules from Campylobacter cells by mass spectrometry and high resolution magic angle spinning NMR spectroscopy. The Journal of Biological Chemistry. 2003;**278**:24509-24520

[66] Karlyshev AV et al. Analysis of Campylobacter jejuni capsular loci reveals multiple mechanisms for the generation of structural diversity and the ability to form complex heptoses. Molecular Microbiology. 2005;**55**:90-103

[67] Gilbert M, Mandrell RE, Parker CT, Li J, Vinogradov E. Structural analysis of the capsular polysaccharide from Campylobacter jejuni RM1221. Chembiochem. 2007;**8**:625-631

[68] McNally DJ et al. The HS:1 serostrain of Campylobacter jejuni has a complex teichoic acid-like capsular polysaccharide with nonstoichiometric fructofuranose branches and O-methyl phosphoramidate groups. The FEBS Journal. 2005;**272**:4407-4422

[69] McNally DJ et al. The HS:19 serostrain of Campylobacter jejuni has a hyaluronic acid-type capsular polysaccharide with a nonstoichiometric sorbose branch and O-methyl phosphoramidate group. The FEBS Journal. 2006;**273**:3975-3989

[70] Bacon DJ et al. A phase-variable capsule is involved in virulence of Campylobacter jejuni 81-176. Molecular Microbiology. 2001;**2001**(40):769-777

[71] Bachtiar BM, Coloe PJ, Fry BN. Knockout mutagenesis of the kpsE gene of Campylobacter jejuni 81116 and its involvement in bacterium–host interactions. FEMS Immunology and Medical Microbiology. 2007;**49**:149-154

[72] Jones MA et al. Adaptation of Campylobacter jejuni NCTC11168 to high-level colonization of the avian gastrointestinal tract. Infection and Immunity. 2004;**72**:3769-3776

[73] Konkel ME, Cieplak W. Altered synthetic response of Campylobacter jejuni to cocultivation with human epithelial cells is associated with enhanced internalization. Infection and Immunity. 1992;**60**:4945-4949

[74] Konkel ME, Kim BJ, Rivera-Amill V, Garvis SG. Bacterial secreted proteins are required for the internalization of Campylobacter jejuni into cultured mammalian cells. Molecular Microbiology. 1999;**32**:691-701

[75] Rivera-Amill V, Kim BJ, Seshu J, Konkel ME. Secretion of the virulence-associated Campylobacter invasion antigens from Campylobacter jejuni requires a stimulatory signal. The Journal of Infectious Diseases. 2001;**183**:1607-1616

[76] Vijayakumar S, Merkx-Jacques A, Ratnayake DB, Gryski I, Obhi RK, Houle S, et al. Cj1121c, a novel UDP-4-keto-6-deoxy-GlcNAc C-4 aminotransferase essential for protein glycosylation and virulence in Campylobacter jejuni. The Journal of Biological Chemistry. 2006;**281**:27733-27743

[77] Fouts DE et al. Major structural differences and novel potential virulence mechanisms from the genomes of multiple Campylobacter species. PLoS Biology. 2005;**3**:e15

[78] Skirrow MB, Blaser MJ. Clinical Aspects of Campylobacter Infection. 2nd ed. Washington, DC: ASM Press; 2000

[79] Gurtler M, Alter T, Kasimir S, Fehlhaber K. The importance of Campylobacter coli in human campylobacteriosis: Prevalence and genetic characterization. Epidemiology and Infection. 2005;**133**:1081-1087. DOI: 10.1017/S0950268805004164 (2005)

[80] Karmali MA, Penner JL, Fleming PC, Williams A, Hennessy JN. The serotype and biotype distribution of clinical isolates of Campylobacter jejuni and Campylobacter coli over a three-year period. The Journal of Infectious Diseases. 1983;**147**:243-246

[81] Kalischuk LD, Inglis GD. Comparative genotypic and pathogenic examination of Campylobacter concisus isolates from diarrheic and non-diarrheic humans. BMC Microbiology. 2011;**11**:53. DOI: 10.1186/1471-2180-11-53

[82] Lindblom GB, Sjogren E, Hansson-Westerberg J, Kaijser B, Campylobacter upsaliensis, C. Sputorum sputorum and C. concisus as common causes of diarrhoea in Swedish children. Scandinavian Journal of Infectious Diseases. 1995;**27**:187-188

[83] O'Donovan D, Corcoran GD, Lucey B, Sleator RD. Campylobacter ureolyticus. Virulence. 2014;**5**:498-506. DOI: 10.4161/viru.28776

[84] Wilkinson DA, O'Donnell AJ, Akhter RN, Fayaz A, Mack HJ, Rogers LE, et al. Updating the genomic taxonomy and epidemiology of Campylobacter hyointestinalis. Scientific Reports. 2018;**8**(1):1-12. DOI: 10.1038/s41598-018-20889-x

[85] On SL. Taxonomy of Campylobacter, Arcobacter, helicobacter and related

bacteria: Current status, future prospects and immediate concerns. Symposium Series Society Application Microbiology. 2001;**90**(S6):1S-15S

[86] Miller WG, Parker CT. Campylobacter and Arcobacter. In: Fratamico P, Liu Y, Kathariou S, editors. Genomes Offoodborne and Waterborne Pathogens. Washington, DC: ASM Press; 2011. pp. 49-65

[87] Blaser MJ. Epidemiologic and clinical features of Campylobacter jejuni infections. The Journal of Infectious Diseases. 1997;**176**(Suppl. 2):S103-S105

[88] Koch-Institut R. RKI-Ratgeber Campylobacter-Enteritis. Epidemiologisches Bulletin. 2018;**23**:1-8

[89] CDC. Foodborne Diseases Active Surveillance Network (FoodNet) Foodborne Diseases Active Surveillance Network (FoodNet). 2019. https://www.cdc.gov/foodnet/reports/cidt-questionsand-answers-2015.html. [Accessed: February 10, 2020]

[90] Colles FM, Maiden MCJ. Campylobacter sequence typing databases: Applications and future prospects. Microbiology. 2012;**158**:2695-2709

[91] CDC Standard Operating Procedure for PulseNet PFGE of *Campylobacter jejuni*. Standard operating procedure for PulseNet PFGE of Campylobacter jejuni. PulseNEt. 2017. https://www.cdc.gov/pulsenet/pathogens/protocols.html. [Accessed: May 14, 2020]

[92] Ayling RD, Woodward MJ, Evans S, Newell DG. Restriction fragment length polymorphism of polymerase chain reaction products applied to the differentiation of poultry campylobacters for epidemiological investigations. Research in Veterinary Science. 1996;**60**:168-172

[93] Melo RT, Grazziotin AL, Valadares Júnior ECV, Prado RR, Mendonca EP, Monteiro GP, et al. Evolution of Campylobacter Jejuni of poultry origin in Brazil. Food Microbiology. 2019;**82**:489-496. DOI: 10.1016/j.fm.2019.03.009

[94] Elhadidy M, Ali MM, El-Shibiny A, Miller WG, Elkhatib WF, Botteldoorn N, et al. Antimicrobial resistance patterns and molecular resistance markers of Campylobacter Jejuni isolates from human diarrheal cases. PLoS One. 2020;**15**(1):227833. DOI: 10.1371/journal.pone.0227833

[95] AWMF, Hagel S, Epple E-J, Feurle GE, Kern WV, Lynnen Jansen P, et al. S2k-Leitlinie Gastrointestinale Infektionen und Morbus Whipple. Stuttgart: Georg Thieme Verlag; 2015

[96] NICE. Diarrhoea and Vomiting Caused by Gastroenteritis in Under 5s Diagnosis and Management. 200. https://www.nice.org.uk/guidance/cg84. [Accessed: April 30, 2020]

[97] Farthing M, Salam MA, Lindberg G, Dite P, Khalif I, Salazar-Lindo E, et al. Acute diarrhea in adults and children: A global perspective. Journal of Clinical Gastroenterology. 2013;**47**(1):12-20. DOI: 10.1097/MCG.0b013e31826df662. PMID: 23222211

[98] World Health Organisation. Critically Important Antimicrobials for Human Medicine 6th Revision. 2018. Available online: https://www.who.int/foodsafety/publications/antimicrobials-sixth/en/. [Accessed: April 1, 2020]

[99] Duarte A, Santos A, Manageiro V, Martins A, Fraqueza MJ, Caniç M, et al. Human, food and animal Campylobacter spp. isolated in Portugal: High genetic diversity and antibiotic resistance rates. International Journal of Antimicrobial Agents. 2014;**44**:306-313 [CrossRef]

[100] European Food Safety Authority; European Centre for Disease Prevention and Control. European union summary report on antimicrobial resistance in zoonotic and indicator bacteria from humans, animals and food in 2016. EFSA Journal. 2018;**16**:4380

[101] European Food Safety Authority; European Centre for Disease Prevention and Control. The European union summary report on antimicrobial resistance in zoonotic and indicator bacteria from humans, aimals and food in 2017. EFSA Journal. 2019;**17**:e04694

[102] Alfredson DA, Korolik V. Antibiotic resistance and resistance mechanisms in Campylobacter jejuni and Campylobacter coli. FEMS Microbiology Letters. 2007;**227**:123-132 [CrossRef]

[103] Bolinger H, Kathariou S. The current state of macrolide resistance in Campylobacter spp.: Trends and impacts of resistance mechanisms. Applied and Environmental Microbiology. 2017;**83**:e00416-e00417 [CrossRef]

[104] Aarestrup FM, Engberg J. Antimicrobial resistance of thermophilic Campylobacter. Veterinary Research. 2001;**32**:311-321 [CrossRef]

[105] Nielsen H, Hansen KK, Gradel KO, Kristensen B, Ejlertsen T, Østergaard C, et al. Bacteraemia as a result of Campylobacter species: A population-based study of epidemiology and clinical risk factors. Clinical Microbiology and Infection. 2010;**16**:57-61 [CrossRef]

[106] Payot S, Bolla J-M, Corcoran D, Fanning S, Mégraud F, Zhang Q. Mechanisms of fluoroquinolone and macrolide resistance in Campylobacter spp. Microbes and Infection. 2006;**8**:1967-1971 [CrossRef]

[107] Drlica K, Zhao X. DNA gyrase, topoisomerase IV, and the 4-quinolones. Microbiology and Molecular Biology Reviews. 1997;**61**:377-392 [CrossRef] [PubMed]

[108] Klostermeier D. Why two? On the role of (A-)symmetry in negative supercoiling of DNA by gyrase. International Journal of Molecular Sciences. 2018;**19**:1489 [CrossRef] [PubMed]

[109] Gellert M, Mizuuchi K, O'Dea MH, Nasht HA. DNA gyrase: An enzyme that introduces superhelical turns into DNA (Escherichia coli/ATP-dependent reaction/superhelix density). Proceedings of the National Academic Science USA. 1976;**73**:3872-3876 [CrossRef] [PubMed]

[110] Zirnstein G, Li Y, Swaminathan B, Angulo F. Ciprofloxacin resistance in Campylobacter jejuni isolates: Detection of gyrA resistance mutations by mismatch amplification mutation assay PCR and DNA sequence analysis. Journal of Clinical Microbiology. 1999;**37**:3276-3280 [CrossRef]

[111] Riddle MS, Guerry P. Status of vaccine research and development for Campylobacter jejuni. Vaccine. 2016;**34**:2903-2906

[112] Ramakrishnan A, Schumack NM, Gariepy CL, Eggleston H, Nunez G, Espinoza N, et al. Enhanced immunogenicity and protective efficacy of a Campylobacter jejuni conjugate vaccine coadministered with liposomes containing monophosphoryl lipid a and QS-21. mSphere. 2019;**4**:e00101-e00119

Chapter 5

Pathobiology, Public Health Significance, and Control of *Campylobacter* Infections

Muhammad Akbar Shahid, Ali Saeed, Sadeeq Ur Rahman, Mian Muhammad Salman, Sheraz Ahmed Bhatti, Muhammad Mudasser Nazir and Muhammad Nauman Zahid

Abstract

Campylobacteriosis is caused by Gram-negative and spiral-shaped microaerophilic *Campylobacter* bacteria. Different avian hosts are commonly infected with *Campylobacter* species. Among 16 *Campylobacter* species, infections are mostly caused by thermophilic *Campylobacter* jejuni and *Campylobacter coli*. *C. jejuni* and *C. coli* are well adapted to the avian intestinal tract and produce little or no clinical diseases in poultry. Although thermophilic *Campylobacters* are commensals in poultry, their significance is due to food safety and public health apprehensions. The majority of human *Campylobacter* infections are caused by *C. jejuni*, followed by *C. coli*, and rarely by *C. lari*. *Campylobacter* infections have now emerged as leading bacterial causes of foodborne gastroenteritis in humans throughout the world. Human Campylobacteriosis cases are sporadic and the disease is characterized by self-limiting watery and/or bloody diarrhea, abdominal pain, and fever; however, severe conditions may occur if patients are immunocompromised. The high prevalence of *Campylobacter* in the intestinal tract of poultry results in contamination of poultry carcasses and poultry products. Handling and eating raw or undercooked poultry meat is considered a significant risk factor for human campylobacteriosis. To ensure food safety and prevent human campylobacteriosis, eradication of *Campylobacter* from the human food chains, especially poultry and poultry products, is indispensable.

Keywords: campylobacteriosis, *Campylobacter* infections, *Campylobacter jejuni*, *Campylobacter coli*, *Campylobacter lari*, poultry-born *Campylobacter*, public health, food safety

1. Introduction

Campylobacteriosis is a bacterial infection affecting both wild and domestic birds and is caused by thermophilic *Campylobacter*. Two important species of genus *Campylobacter*, that is, *C. jejuni* and *C. coli* are responsible for producing disease in birds. *Campylobacter* species are gram-negative rod-shaped bacteria. *Campylobacter* is an enteric organism that inhabits the intestinal tract of birds and is excreted through feces [1]. The disease is primarily spread horizontally, and vertical transmission is thought to be quite uncommon. It is a significant zoonotic infection that causes diarrheal sickness in humans when consumed through tainted meat, food, and water. Human intestinal campylobacteriosis symptoms include fever, diarrhea, and stomach pain. The clinical course of enteritis usually resolves on its own, but some infected people experience severe post-infectious complications such as autoimmune diseases that affect the brain system, joints, and intestinal tract. Moreover, systemic pathogen spread in immunocompromised people might result in circulatory disorders and septicemia [1].

As a pathogen, *Campylobacter* is extremely important for both public health and food safety. Human campylobacteriosis is typically caused by *C. jejuni*, followed by *C. coli*, and less frequently, *C. lari*. Treatment failures in human patients have been caused by the development of resistance in many avian *Campylobacter* isolates to antibiotics such as macrolides and fluoroquinolones [2]. *Campylobacter* infections are estimated to cost society several billion dollars yearly in socioeconomic expenses.

Major human infection sources include poultry meat products. To lessen the burden of campylobacteriosis, public health authorities, veterinarians, doctors, researchers, and legislators must work together under the guiding principle of "One World—One Health" [3, 4]. Improvements in information dissemination to strengthen hygiene measures for agricultural remediation are among the innovative intervention regimes for the prevention of *Campylobacter* contaminations along the food chain. Novel intervention tactics strengthen both the decrease of pathogen contamination in food production and the treatment of the related disorders in people because it is not possible to completely eradicate *Campylobacter* from the food production chains [2].

2. Epidemiology of *Campylobacter* infections

2.1 Incidence and distribution

Campylobacteriosis has been reported to be prevalent in both domestic and wild birds but the former is considered to be more affected [5]. The possible reason might be that transmission among domestic birds, especially commercial poultry, is high due to more number of birds in a unit area. Different factors in commercial farming may affect the occurrence of campylobacteriosis, including the type of farming, housing system, region, and biosecurity measures. It has also been reported that the prevalence of campylobacteriosis is high in months in which the temperature is high, resulting in a higher population of flies and higher flies-mediated transmission [6]. Developing countries and European countries are considered to have a high prevalence of the disease as compared to Scandinavian countries [7]. The age of the birds, irrespective of the species and production system, is related to the occurrence of the disease and it being less likely to occur in birds of less than 2–3 weeks of age. The occurrence of *C. jejuni* isolates among other *Campylobacter* species is high followed by *C. coli* and *C. lari* [8]. The isolation of other species of *Campylobacter*, including *C. upsaliensis* and *C. hyointestinalis*, is very low from poultry [1, 5].

2.2 Transmission, carriers, and vectors

Horizontal transmission is the most common mode by which transmission of campylobacteriosis takes place. Vertical transmission does not occur or occurs very rarely. The possible sources for horizontal transmission from the environment to poultry include contaminated water [9], litter especially old litter [10], farm workers [11], contaminated footwear, insects [12], wild animals [13] especially rodents [14], farm animals [15], scavenger birds [16], feed contaminated with feces of chicks [15], house flies [17], visitors, and various types of equipment.

Campylobacter is usually excreted through feces, which may contaminate the feed and litter. Survival of *Campylobacter* in a litter depends upon temperature, moisture, and pH [18, 19], although *Campylobacter* can survive in the litter for a minimum of 10 days at 20°C [20]. An infected water supply may also result in the spreading of the disease among the flocks. Insects such as houseflies, darkling beetles, cockroaches, and mealworms play an important role as a mechanical vector in the transmission of *Campylobacters* [21, 22].

3. Pathobiology of *Campylobacter* infections

3.1 Incubation period

The incubation period of *Campylobacter* ranges from 2 to 5 days in avian species. Birds can be infected by *Campylobacter* naturally, and experimental infection of *Campylobacter* can also be produced in birds but mostly the clinical signs are not visible. The appearance of clinical signs especially gastrointestinal signs is related to the age of the host. Chickens experimentally infected with *Campylobacter* on the first day, 12 hours post-hatching, resulted in the appearance of diarrhea while no clinical manifestation of campylobacteriosis was observed in 3-day-old chicks that were infected with 10^9 organisms [23]. In a day-old chick, a dose of as low as 2 cfu, has been established for *Campylobacter* colonization [24]. Chickens aged between 2 and 3 weeks of age, reared at commercial farms, have not been shown to be infected by *Campylobacter* infection and it may be associated with the presence of maternal antibodies [25, 26]. Flocks infected with *Campylobacter* specifically *C. jejuni* will shed the organism for at least 12 weeks of age [27]. The shedding may continue for 42 weeks in the breeder birds [28]. The incubation period of *Campylobacters* in humans is from 2 to 4 days but can range from 1 to 10 days [29].

3.2 Clinical signs and pathological lesions

Under natural conditions, clinical signs are not observed in poultry infected with *Campylobacter*. Clinical signs, including weight loss and diarrhea, have been reported in young birds that have been experimentally challenged with *Campylobacter* infection [30], and diarrhea may last for 7–14 days. The gastrointestinal tract was reported to be the site where minimal microscopic and pathological lesions were observed in experimentally infected birds [31]. Gross lesions observed in chicks infected by *Campylobacter* included mucus and fluid-filled distended jejunum [32] and petechial hemorrhages of the mucosa [23]. The microscopic lesions include edema of mucosa and submucosa of the GIT, especially in the cecum [23] and *Campylobacter* may be found attached to the brush border of enterocytes [32]. In severe conditions, the intestinal lumen may be filled with erythrocytes and leukocytes due to mononuclear infiltration of the submucosa and villous atrophy [23].

Consumption of raw milk, non-chlorinated or contaminated surface water, and ingestion of raw or undercooked poultry or red meat are some of the ways that humans might contract *Campylobacter* infections. Close contact with sick pets in a household setting can also result in human *Campylobacter* infections [33]. *Shigella* and *Salmonella* infections can sometimes be difficult to distinguish clinically from *Campylobacter* infections [34]. The mechanisms of *Campylobacter* survival and infection are poorly known, but when it colonizes the ileum, jejunum, and colon, it can occasionally result in infection with or without symptoms. The transmission cycle of *Campylobacter* infections is shown in **Figure 1**.

Human gastroenteritis is frequently brought on by *Campylobacter*, however, the infection can also develop beyond the intestines. Two forms of *Campylobacter* infections exist, that is, gastrointestinal infection (GI) and extragastrointestinal infection (EI). Diarrhea is typically a symptom of gastroenteritis, an inflammation of the gastrointestinal tract that affects the small intestine and stomach. *Campylobacter* is one of the four major bacterial causes of gastrointestinal illnesses worldwide [35]. Moreover, it is a significant and frequent cause of children's diarrhea and traveler's diarrhea [36]. Reactive arthritis, Guillain-Barre syndrome (GBS) [37], bacteremia, septicemia, septic arthritis, endocarditis, neonatal sepsis, osteomyelitis, and meningitis are among the extragastrointestinal illnesses linked to *Campylobacter* infections [1]. Other extragastrointestinal post-infections linked to Campylobacter infections include severe neurological dysfunction, neurological abnormalities, and paralysis resembling polio in a rare number of patients [35].

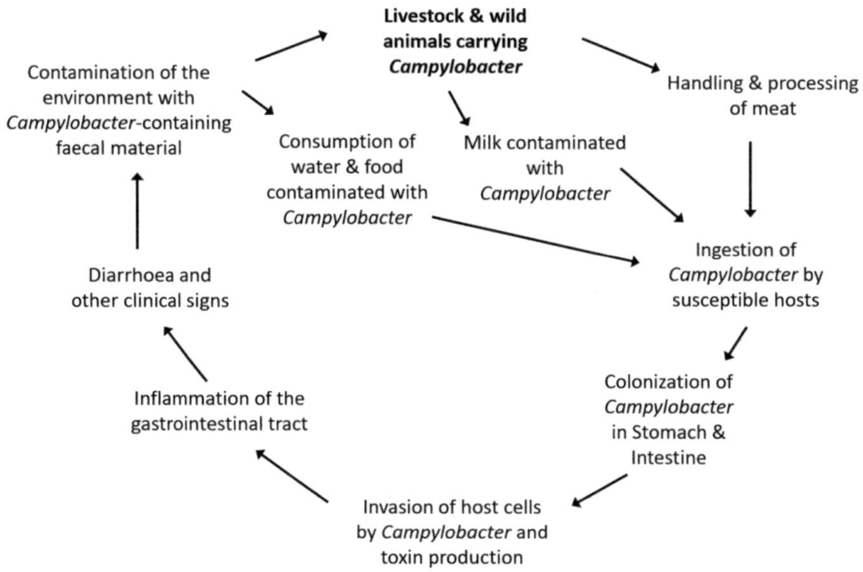

Figure 1.
The transmission cycle of Campylobacter infections.

DOI: http://dx.doi.org/10.5772/intechopen.112216

3.3 Pathogenesis of the *Campylobacter* infections

The ability of *Campylobacter* to survive outside the gut is very low and it does not replicate outside the gut [38] and temperature ranging from 37 to 42°C is suitable for its growth. Thus, a chicken's body temperature (41–42°C) is suitable for the growth and survival of *Campylobacter* [39]. *Campylobacter* species gain entry into the body of the bird *via* the fecal-oral route and colonize the caecum, cloaca, and distal jejunum [40]. The most probable site for the colonization of *Campylobacter* are the crypts of the cloaca and cecum but it may be found in minute levels in the small intestine and gizzard. The colonization of *Campylobacter* in the intestine of birds is affected by various factors [41]. Colonization of intestinal epithelium is accomplished by chemotaxis with the help of chemoattractants, including mucin and L-Fucose [42]. Flagellum helps the organism in its movement in a viscous fluid and helps it in colonizing the intestinal mucosa. Different outer membrane proteins and LPS have been associated with adhesion and invasion. After colonizing the intestinal epithelium of the intestine, CLT (cholera-like-toxin) and cytotoxin result in tissue damage leading to inflammation followed by leakage of serosal fluid [38].

4. Public health significance of campylobacteriosis

Campylobacteriosis is continuously a serious public health concern, especially in developing countries. The incidence of campylobacteriosis is substantially increased in the last couple of decades with a high morbidity rate and significant infant mortality. Moreover, emerging new species and antibiotic resistance in most common species, including *Campylobacter jejuni* are additional challenges in the control of *Campylobacter* infections [43]. A serious methodological effort is required for public awareness and disease control with the involvement of all stakeholders. Primarily, a continuous ongoing surveillance program is required with proper laboratory infrastructure for the diagnosis along with fundamental and effective enteric disease control programs, especially in developing countries. Further, a systematic approach is required to control *Campylobacter* infections, including proper monitoring of disease burden, source attribution, risk assessment and management, surveillance of antimicrobial resistance, and assessment of possible control measures. However, thermophilic *Campylobacter* is ubiquitously present, but most recent outbreaks were commonly associated with water and food cross-contamination with animal shedding. Although, animals are asymptomatic carriers of *Campylobacter*, cross-contamination of the food chain with animal waste at the different stages of slaughtering, processing and marketing, direct human contact with pets, and contamination of drinking water with animal excreta possibly lead to disease outbreaks in human [3]. Indeed, *Campylobacter* spp. and sources of food chain contaminations should also be taken into account while developing disease control strategies.

4.1 Burden of the disease and risk assessment

The disease burden is difficult to predict in the case of campylobacteriosis. Population-based cohort studies are commonly used to estimate the disease burden, especially in developed countries. According to the two population-based cohorts, the incidence of gastroenteritis due to *Campylobacter* spp. was one out of seven and one out of four people in the UK and Netherlands, respectively [43]. Cohort-based studies are more common than population-based studies.

Campylobacteriosis accounts for 7.5 million DALY (disability-adjusted life years) or 8.4% of the global burden of diarrheal diseases, according to the Global Burden of Disease (GBD) project, and ranks fourth among identified pathogens after rotavirus (18.7 million DALY), typhoid fever (12.2 million DALY), and cryptosporidiosis (8.3 million DALY) [44]. Estimating disease burden enables the implementation of potential biosafety or control measures as well as the evaluation of the disease and/or outbreak situation in specific population areas.

4.2 Source attribution and risk assessment of *Campylobacter* infections

The disease source and transmission routes are also assessed with microbial source attributions. *Campylobacter* spp. are isolated from human infections, and gene sequences are compared with *Campylobacter* spp. isolated from food and environmental sources. Similarly, a multilocus sequence typing (MLST) is used for the source attribution in the epidemiological investigation of rural and urban populations of New Zealand [45] and the United Kingdom [46], which indicated that both populations have different epidemiological patterns of *Campylobacter*. The cost of annual disease attribution is still very high. According to an estimate, the annual attribution cost of *Campylobacter* infections among other diarrhoeal diseases in the USA is approximately 1.2–4 billion USD per year [47, 48] and 2.4 billion EUR in the EU [49]. Therefore, the estimation of disease burden is constantly becoming an important parameter to assess disease risk and to develop an effective health care policy. However, a unified and consistent risk assessment plan is extremely desirable in campylobacteriosis. Previously, cross-contamination of the food chain with *Campylobacter* from poultry carcass has been successfully estimated with two mathematical risk assessment models [50], which indicates that human incidence of Campylobacteriosis is reduced up to 30 times with 2 log reduction of the *Campylobacter* number on poultry carcass. However, a unified accurate quantitative risk assessment model is difficult to develop in this disease due to the continuous emergence of genetic variation in *Campylobacter* spp., the subsequent diverse range of virulence, and different host-immune defenses [3]. However, several previous studies indicate that cross-contamination of the food chain from poultry carcasses is the most common source of human infections of *Campylobacter*. The control of poultry-born *Campylobacter* cross-contamination in the food chain can be one of the potential control measures to reduce the human incidence of campylobacteriosis.

4.3 Risk management and control measures

The risk management plans should be implemented according to the species of *Campylobacter* infections, disease source, and risk assessment recommendations. The reduction of bacterial numbers in poultry carcasses, the most common cause of campylobacteriosis, can be achieved with strict biosecurity measures in poultry flocks, appropriate slaughtering procedures, and hygienic meat processing methods. On other hand, bacteriophage can also be used to reduce pathogenic bacterial numbers in food chains. Previously, the reduction of bacterial count in the food chain up to the magnitude of two has been successfully achieved by the application of bacteriophage [51]. Additionally, physical and chemical decontamination and disinfection methods can also reduce the bacterial number in poultry carcasses and subsequently incidence of campylobacteriosis. Further, the continuous surveillance of *Campylobacter* spp. against different antimicrobial agents should also be monitored and the impact of antimicrobial use should also be regularly assessed in the risk assessment and disease attribution process [52]. Public health authorities should also introduce public awareness programs about different sources of *Campylobacter* infections, health impacts, and possible control or safety measures. The increasing incidence of campylobacteriosis in developing countries [53] further indicates the need for accurate disease surveillance along with strict food safety regulations followed by alleviation strategies to control *Campylobacter* infections in these areas. The economic burden to implement mitigation strategies is an additional hitch in the control of campylobacteriosis in these countries. However, a hygienic food chain supply, safe contact with pet animals, and public awareness program about *Campylobacter* infections can collectively improve the epidemiological prevalence and public health in developing countries.

5. Conclusion

It is concluded that consuming contaminated meat, food, and water cause human *Campylobacter* illnesses. Since the organism is zoonotic, veterinary and human medicine must work together under the "one-health" tenet to develop efficient and better methods for preventing and controlling infections in both human and animal populations. For a better understanding of the origins of infections, DNA-based investigations should be employed to ascertain the genetic relatedness of human and animal C*ampylobacters* for developing better prevention and control strategies. Continuing research and surveillance are required to better understand the patterns and trends of antibiotic resistance in *Campylobacter* isolates collected from both humans and animals. Exploration of environmental *Campylobacter* reservoirs and related risk factors for human and animal infections with *Campylobacter* is also necessary.

Conflict of interest

The authors declare no conflict of interest.

Author details

Muhammad Akbar Shahid[1*], Ali Saeed[2,3], Sadeeq Ur Rahman[4,5], Mian Muhammad Salman[4,5], Sheraz Ahmed Bhatti[1], Muhammad Mudasser Nazir[6] and Muhammad Nauman Zahid[7]

1 Department of Pathobiology, Faculty of Veterinary Sciences, Bahauddin Zakariya University, Multan, Pakistan

2 Institute of Molecular Biology and Biotechnology, Bahauddin Zakariya University, Multan, Pakistan

3 Department of Pediatric Oncology and Medical Microbiology, University Medical Center Groningen, University of Groningen, Groningen, The Netherlands

4 College of Veterinary Sciences and Animal Husbandry, Abdul Wali Khan University, Mardan, Pakistan

5 Department of Microbiology, Timergara Campus, Abdul Wali Khan University, Mardan, Pakistan

6 Faculty of Veterinary and Animal Sciences, Department of Parasitology, The Islamia University of Bahawalpur, Bahawalpur, Pakistan

7 Department of Biology, College of Science, University of Bahrain, Kingdom of Bahrain

*Address all correspondence to: drakbarshahid@gmail.com; makbar@bzu.edu.pk

References

[1] Igwaran A, Okoh AI. Human campylobacteriosis: A public health concern of global importance. Heliyon. 2019;**5**(11):e02814

[2] Heimesaat MM, Backert S, Alter T, Bereswill S. Human Campylobacteriosis-a serious infectious threat in a one health perspective. Current Topics in Microbiology and Immunology. 2021;**431**:1-23

[3] Gölz G, Rosner B, Hofreuter D, Josenhans C, Kreienbrock L, Löwenstein A, et al. Relevance of Campylobacter to public health--the need for a one health approach. International Journal of Medical Microbiology. 2014;**304**(7):817-823

[4] Heimesaat MM, Backert S, Alter T, Bereswill S. Human Campylobacteriosis—A serious infectious threat in a one health perspective. In: Backert S, editor. Fighting Campylobacter Infections: Towards a One Health Approach. Cham: Springer International Publishing; 2021. pp. 1-23

[5] Shane SM. The significance of Campylobacter jejuni infection in poultry: A review. Avian Pathology. 1992;**21**(2):189-213

[6] Hald B, Skovgård H, Bang DD, Pedersen K, Dybdahl J, Jespersen JB, et al. Flies and Campylobacter infection of broiler flocks. Emerging Infectious Diseases. 2004;**10**(8):1490-1492

[7] Newell DG, Fearnley C. Sources of Campylobacter colonization in broiler chickens. Applied and Environmental Microbiology. 2003;**69**(8):4343-4351

[8] Nielsen EM, Engberg J, Madsen M. Distribution of serotypes of Campylobacter jejuni and C. coli from Danish patients, poultry, cattle and swine. FEMS Immunology and Medical Microbiology. 1997;**19**(1):47-56

[9] Pearson AD, Greenwood M, Healing TD, Rollins D, Shahamat M, Donaldson J, et al. Colonization of broiler chickens by waterborne Campylobacter jejuni. Applied and Environmental Microbiology. 1993;**59**(4):987-996

[10] Montrose MS, Shane SM, Harrington KS. Role of litter in the transmission of Campylobacter jejuni. Avian Diseases. 1985;**29**(2):392-399

[11] Berndtson E, Emanuelson U, Engvall A, Danielsson-Tham ML. A 1-year epidemiological study of campylobacters in 18 Swedish chicken farms. Preventive Veterinary Medicine. 1996;**26**(3):167-185

[12] Ekdahl K, Normann B, Andersson Y. Could flies explain the elusive epidemiology of campylobacteriosis? BMC Infectious Diseases. 2005;**5**:11

[13] Lillehaug A, Monceyron Jonassen C, Bergsjø B, Hofshagen M, Tharaldsen J, Nesse LL, et al. Screening of feral pigeon (Colomba livia), mallard (Anas platyrhynchos) and graylag goose (Anser anser) populations for Campylobacter spp., salmonella spp., avian influenza virus and avian paramyxovirus. Acta Veterinaria Scandinavica. 2005;**46**(4):193-202

[14] Petersen L, Wedderkopp A. Evidence that certain clones of Campylobacter jejuni persist during successive broiler flock rotations. Applied

and Environmental Microbiology. 2001;**67**(6):2739-2745

[15] Gregory E, Barnhart H, Dreesen DW, Stern NJ, Corn JL. Epidemiological study of Campylobacter spp. in broilers: Source, time of colonization, and prevalence. Avian Diseases. 1997;**41**(4):890-898

[16] Craven SE, Stern NJ, Line E, Bailey JS, Cox NA, Fedorka-Cray P. Determination of the incidence of salmonella spp., Campylobacter jejuni, and Clostridium perfringens in wild birds near broiler chicken houses by sampling intestinal droppings. Avian Diseases. 2000;**44**(3):715-720

[17] Rosef O, Kapperud G. House flies (Musca domestica) as possible vectors of Campylobacter fetus subsp. jejuni. Applied and Environmental Microbiology. 1983;**45**(2):381-383

[18] Doyle MP, Roman DJ. Sensitivity of Campylobacter jejuni to drying. Journal of Food Protection. 1982;**45**(6):507-510

[19] Smitherman RE, Genigeorgis CA, Farver TB. Preliminary observations on the occurrence of Campylobacter jejuni at four California chicken ranches. Journal of Food Protection. 1984;**47**(4):293-298

[20] Luechtefeld NW, Wang WL, Blaser MJ, Reller LB. Evaluation of transport and storage techniques for isolation of Campylobacter fetus subsp. jejuni from Turkey cecal specimens. Journal of Clinical Microbiology. 1981;**13**(3):438-443

[21] Rollins DM, Colwell RR, Pearson AD. Non-culturable *campylobacters* in the environmment: Detection by indirect fluorescent antibody technique. In: Kaijser B, Falsen E, editors. *Campylobacter IV*. Göteborg, Sweden: University of Gothenburg; 1987. pp. 285-291

[22] Rollins DM. Potential for reduction in colonization of poultry by Campylobacter from environmental sources. In: Blankenship LC, Bailey JS, Cox NA, Craven SE, Meinersmann RJ, Stern NJ, editors. Colonization Control of Human Bacterial Enteropathogens in Poultry. San Diego, California: Academic Press; 1991. p. 47-56

[23] Welkos SL. Experimental gastroenteritis in newly-hatched chicks infected with Campylobacter jejuni. Journal of Medical Microbiology. 1984;**18**(2):233-248

[24] Knudsen KN, Bang DD, Andresen LO, Madsen M. Campylobacter jejuni strains of human and chicken origin are invasive in chickens after oral challenge. Avian Diseases. 2006;**50**(1):10-14

[25] Sahin O, Zhang Q, Meitzler JC, Harr BS, Morishita TY, Mohan R. Prevalence, antigenic specificity, and bactericidal activity of poultry anti-Campylobacter maternal antibodies. Applied and Environmental Microbiology. 2001;**67**(9):3951-3957

[26] Sahin O, Luo N, Huang S, Zhang Q. Effect of Campylobacter-specific maternal antibodies on Campylobacter jejuni colonization in young chickens. Applied and Environmental Microbiology. 2003;**69**(9):5372-5379

[27] Kaino K, Hayashidani H, Kaneko K, Ogawa M. Intestinal colonization of Campylobacter jejuni in chickens. Nihon juigaku zasshi The Japanese journal of veterinary science. 1988;**50**(2):489-494

[28] Lindblom GB, Sjörgren E, Kaijser B. Natural campylobacter colonization

in chickens raised under different environmental conditions. The Journal of Hygiene (Lond). 1986;**96**(3):385-391

[29] Mark EL, Kevin C, Aimee LG. Campylobacteriosis: National Center for Emerging and Zoonotic Infectious Diseases (NCEZID), Division of Global Migration and Quarantine (DGMQ). 2019. Available from: https://wwwnc.cdc.gov/travel/yellowbook/2020/travel-related-infectious-diseases/campylobacteriosis.

[30] Ruiz-Palacios GM, Escamilla E, Torres N. Experimental Campylobacter diarrhea in chickens. Infection and Immunity. 1981;**34**(1):250-255

[31] Zhang Q, Sahin O. Campylobacteriosis. In: Diseases of Poultry. 13th ed. Ames, IA: John Wiley & Sons, Inc.; 2013. pp. 737-750

[32] Sanyal SC, Islam KM, Neogy PK, Islam M, Speelman P, Huq MI. Campylobacter jejuni diarrhea model in infant chickens. Infection and Immunity. 1984;**43**(3):931-936

[33] Shane SM. Campylobacter infection of commercial poultry. Revue Scientifique et Technique. 2000;**19**(2):376-395

[34] Hansson I, Sandberg M, Habib I, Lowman R, Engvall EO. Knowledge gaps in control of Campylobacter for prevention of campylobacteriosis. Transboundary and Emerging Diseases. 2018;**65**(Suppl. 1):30-48

[35] WHO. WHO estimates of the global burden of foodborne diseases: Foodborne disease burden epidemiology reference group 2007-2015. Geneva: World Health Organization; 2015

[36] Liu J, Platts-Mills JA, Juma J, Kabir F, Nkeze J, Okoi C, et al. Use of quantitative molecular diagnostic methods to identify causes of diarrhoea in children: A reanalysis of the GEMS case-control study. Lancet. 2016;**388**(10051):1291-1301

[37] Kuwabara S, Yuki N. Axonal Guillain-Barré syndrome: concepts and controversies. Lancet Neurology. 2013;**12**(12):1180-1188

[38] Ketley JM. Pathogenesis of enteric infection by Campylobacter. Microbiology (Reading, England). 1997;**143**(Pt 1):5-21

[39] Campylobacter NI, Arcobacter. Manual of Clinical Microbiology. 6th ed. Washington: ASM Press; 1995. pp. 483-491

[40] Beery JT, Hugdahl MB, Doyle MP. Colonization of gastrointestinal tracts of chicks by Campylobacter jejuni. Applied and Environmental Microbiology. 1988;**54**(10):2365-2370

[41] Doyle MP. Association of Campylobacter jejuni with laying hens and eggs. Applied and Environmental Microbiology. 1984;**47**(3):533-536

[42] Hugdahl MB, Beery JT, Doyle MP. Chemotactic behavior of Campylobacter jejuni. Infection and Immunity. 1988;**56**(6):1560-1566

[43] World Health O. The Increasing Incidence of Human Campylobacteriosis Report and Proceedings of a WHO Consultation of Experts, Copenhagen, Denmark, 21-25 November 2000. Geneva: World Health Organization; 2001

[44] Murray CJ, Vos T, Lozano R, Naghavi M, Flaxman AD, Michaud C, et al. Disability-adjusted life years (DALYs) for 291 diseases and injuries in 21 regions, 1990-2010: A systematic analysis for the global burden of disease study 2010. Lancet. 2012;**380**(9859):2197-2223

[45] Mullner P, Shadbolt T, Collins-Emerson JM, Midwinter AC, Spencer SE, Marshall J, et al. Molecular and spatial epidemiology of human campylobacteriosis: Source association and genotype-related risk factors. Epidemiology and Infection. 2010;**138**(10):1372-1383

[46] Strachan NJ, Gormley FJ, Rotariu O, Ogden ID, Miller G, Dunn GM, et al. Attribution of Campylobacter infections in Northeast Scotland to specific sources by use of multilocus sequence typing. The Journal of Infectious Diseases. 2009;**199**(8):1205-1208

[47] Eberle KN, Kiess AS. Phenotypic and genotypic methods for typing Campylobacter jejuni and Campylobacter coli in poultry. Poultry Science. 2012;**91**(1):255-264

[48] Batz M, Hoffmann S, Morris JG Jr. Disease-outcome trees, EQ-5D scores, and estimated annual losses of quality-adjusted life years (QALYs) for 14 foodborne pathogens in the United States. Foodborne Pathogens and Disease. 2014;**11**(5):395-402

[49] Anonymous. EFSA Explains Zoonotic Diseases: Campylobacter | European Food Safety Authority. 2014. https://www.efsa.europa.eu/en/corporate/pub/factsheetcampylobacter. [Accessed: December 15, 2023]

[50] Rosenquist H, Nielsen NL, Sommer HM, Nørrung B, Christensen BB. Quantitative risk assessment of human campylobacteriosis associated with thermophilic Campylobacter species in chickens. International Journal of Food Microbiology. 2003;**83**(1):87-103

[51] Hertwig S, Hammerl JA, Appel B, Alter T. Post-harvest application of lytic bacteriophages for biocontrol of foodborne pathogens and spoilage bacteria. Berliner und Münchener Tierärztliche Wochenschrift. 2013;**126**(9-10):357-369

[52] Aarestrup FM. The livestock reservoir for antimicrobial resistance: A personal view on changing patterns of risks, effects of interventions and the way forward. Philosophical Transactions of the Royal Society of London. Series B, Biological Sciences. 2015;**370**(1670):20140085

[53] Premarathne J, Satharasinghe DA, Huat JTY, Basri DF, Rukayadi Y, Nakaguchi Y, et al. Impact of human Campylobacter infections in Southeast Asia: The contribution of the poultry sector. Critical Reviews in Food Science and Nutrition. 2017;**57**(18):3971-3986

Chapter 6

Quorum Sensing in Biofilm

Zahra Sedarat and Andrew W. Taylor-Robinson

Abstract

Quorum sensing (QS) is a complex system of communication used by bacteria, including several notable pathogens that pose a significant threat to public health. The central role of QS in biofilm activity has been demonstrated extensively. The small extracellular signaling molecules, known as autoinducers, that are released during this process of cell-to-cell communication play a key part in gene regulation. QS is involved in such diverse intracellular operations as modulation of cellular function, genetic material transfer, and metabolite synthesis. There are three main types of QS in bacteria, metabolites of which may form the target for novel treatment approaches. The autoinducing peptide system exists only in Gram-positive bacteria, being replaced in Gram-negative species by the acyl-homoserine lactone system, whereas the autoinducer-2 system occurs in both.

Keywords: bacterium, gram-positive, gram-negative, biofilm, quorum sensing, quorum quenching, autoinducer, accessory gene regulator, acyl-homoserine lactone, LuxS, luminescence, *Staphylococcus aureus*, *Vibrio fischeri*, *Vibrio harveyi*

1. Introduction

More than half a century ago, pioneering experiments performed on *Streptococcus pneumoniae* discovered the existence of bacterial communication through hormone-like molecules that were later defined as peptides [1, 2]. These findings were accomplished by studying *Vibrio fischeri*, which is able to produce luminescence at high levels of cell density. The luminous system in this marine bacterium is characteristically self-regulated and is provoked at a threshold level of signal molecules. This so-called "autoinduction" provides an environmental sensing mechanism [3]. Known as "quorum sensing," this type of regulation is a form of information sharing used by bacteria through intercellular communication to regulate gene expression. This process is described in many Gram-positive and Gram-negative bacteria. It is facilitated via autoinducers (AIs) or extracellular signaling molecules that produce, release, and detect as well as respond to them [4]. By increasing bacterial cell density, accumulated AIs in the outer cell will lead to changes in gene expression. This communication is detected in inner species and also between species. Among multiple cell activities that are under the control of QS, biofilm formation, virulence factor formation, sporulation, motility, conjugation, symbiosis, competence, and sporulation are each of note [5–8]. In addition, a number of studies have shown the key role of quorum sensing in metabolic processes, involving a high portion of the bacterial genome (corresponding to more than 20% of the proteome) that facilitates adaptation to metabolic needs [9].

The bacteria belonging to the same colony may exhibit heterogenous phenotypic behavior in order to respond to environmental fluctuations and interbacterial interactions. These are coordinated with each other via quorum sensing, which adapts bacterial traits and behaviors (both group and individual) to ensure their compatibility [10]. In short, QS plays a fundamental role in production, detection, and response to AIs [11].

In the QS system used by various bacteria, there are differences in terms of target genes, types of chemical signal molecules, and mechanisms [8]. Emerging evidence points to several types of signaling molecules, including methyl dodecanoic acid, N-acyl homoserine lactones (AHLs), furanosyl borate, oligopeptides, and hydroxy palmitic acid methyl ester [12]. Although there are multiple QS systems described in bacteria, these are broadly categorized into three groups that we will describe in detail in this chapter. The first major group belongs to Gram-negative bacteria and uses AHLs as the signaling molecule [6]. The second group, only found in Gram-positive bacteria, utilizes small, processed oligopeptides [8]. The third group, in which autoinducer-2 (AI-2) is produced, applies to both Gram-positive and Gram-negative bacteria and has been reported in over 55 species [13].

1.1 Quorum sensing in Gram-negative bacteria

Some characteristics of QS are common to Gram-negative bacteria. The main feature is the ability of AHLs and s-adenosylmethionine-synthetized molecules to diffuse within the bacterial membrane. The receptors for these are located either in the cytoplasm or on the inner membrane. Additionally, numerous cell processes are affected by QS, which directly modifies the relevant genes [14, 15]. Different types of autoinducers are used by Gram-negative bacteria, whereas the most common type, Acyl-HSL, is found in many bacterial species [14, 16, 17]. The AHLs (lux operon) were first described in *Vibrio fischeri*, which will be discussed as a model in this section [18]. An important reason why *V. fischeri* QS is suitable to study is its high sensitivity to AIs, which means it is activated even when they are at low levels [19–21].

In general, AHL-mediated QS involves either LuxI or LuxR proteins [22]. These are engaged in multiple cell functions including biofilm formation, pathogenesis, antibiotic production, and genetic competence. Hence, LuxI-LuxR is considered an excellent research model [23]. Indeed, the operon LuuxICDABEG is activated by LuxR [22]. More than 20 LuxR analogous families exist in Gram-negative bacteria, of which LuxR is the most studied [24]. LasI and EsaI in *Pseudomonas aeruginosa* and *Pantoea stewartii*, respectively, are of note [25, 26].

LuxR should first be activated by the AIs, N-(3-oxohexanoyl)-L-homoserine lactone (abbreviated to 3-oxo-C6-HSL). This is a diffusible signal catalyzed by a 193-amino acid protein that is encoded by LuxI from a precursor of host metabolism (s-adenosyl methionine) as well as a cofactor acyl carrier protein. In addition to 3-oxo-C6, the other products of LuxI, are apo-ACP and 5′-methylthioadenosine [8, 22, 24, 27, 28]. Thus, in the presence of AI (3-oxo-C6), LuxR activates LuxICDABEG operon expression, and overexpression of LuxR will be followed too [29]. The C-terminal region of LuxR is responsible for DNA-binding as well as RNA polymerase interaction (resulting in activation of the Lux promotor), whereas the N-terminal binds to AIs [30–32].

Other parts of the Lux operon are associated with diverse activities. LuxAB is in charge of encoding luciferase (a heterodimer of two subunits, alpha and beta). LuxC, LuxD, and LuxE are responsible for encoding aldehyde substrate, whereas LuxG regenerates FMNH2 from FMN [24, 33, 34]. In this regard, luciferase and flavin-dependent monooxygenase, which produce light photons from chemical energy via catalyzing a bioluminescent reaction, facilitate an enzymatic reaction to produce aliphatic acid (RCOOH) as well as FMN from substrates including FMNH2, O2, and long-chain fatty acids (RCHO). In this way, bacteria regulate luminescence production in light organs of fish at high cell density and switch on *lux* genes (**Figure 1**) [34–37].

Lastly, an intergenic region known as Lux box (a 20-bp palindromic sequence) is located inside the LuxI promoter within 42.5 bp of the LuxICDABEG operon start site. This acts as a transcriptional activator that is responsible for the overexpression of the LuxI promotor [38–40]. Although the Lux box plays an essential part in luminescence gene activation, its precise role and structure remain to be identified [39].

1.2 Quorum sensing in Gram-positive bacteria

Autoinduction by Gram-positive bacteria is achieved via autoinducer peptides (AIPs) that require postproduction processing. AIPs are not permeable and require carriage across the host cell membrane by transporter proteins [41–43]. Additionally, two types of transcription factors are recognized, Rgg and RNPP, the latter of which is found in all Gram-positive bacteria and is equipped with a binding domain that facilitates its binding to signaling peptides [44].

In the model bacterium *Staphylococcus aureus* (**Figure 2**), there are four types of two-component regulator system, namely, *agr*AC, *sae*RS, *arl*RS, and *srr*AB. Of these,

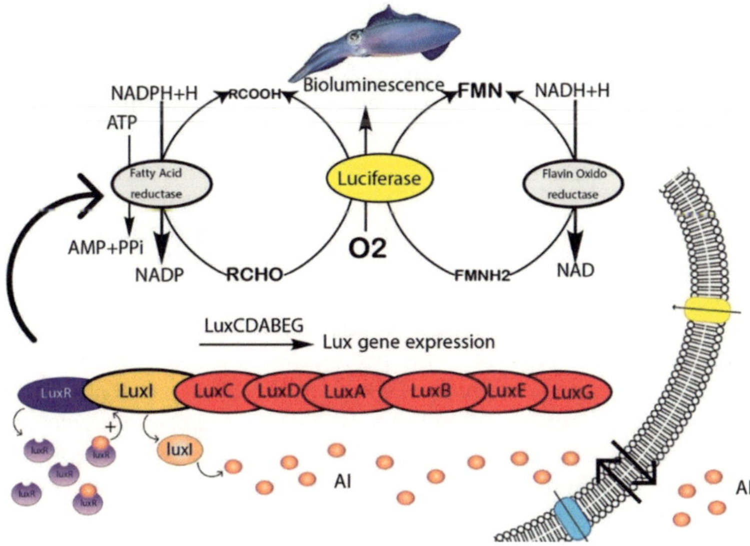

Figure 1.
Lux quorum sensing system in Vibrio fischeri. Autoinducers (AIs) are synthetized by LuxI which later attaches to the LuxR protein at threshold concentration. LuxR protein acts as receptor and its complex with AIs raises LuxI gene expression and thus AI production. AIs diffuse through the cell membrane and thereby activate LuxR protein. LuxCDABEG encodes the structural components of light production in which LuxAB encodes luciferase. Bioluminescence and light production happen after oxidation of RCHO and FMNH2. Production of fatty acids and activation of fatty acyl groups are functions performed by LuxD and LuxC, respectively. Activated fatty acyl groups are then reduced to long-chain aldehydes by LuxE. Recycling of components such as fatty acids as well as FMN, which is produced as a result of the luciferase reaction, are carried out by LuxC/E and LuxG, respectively.

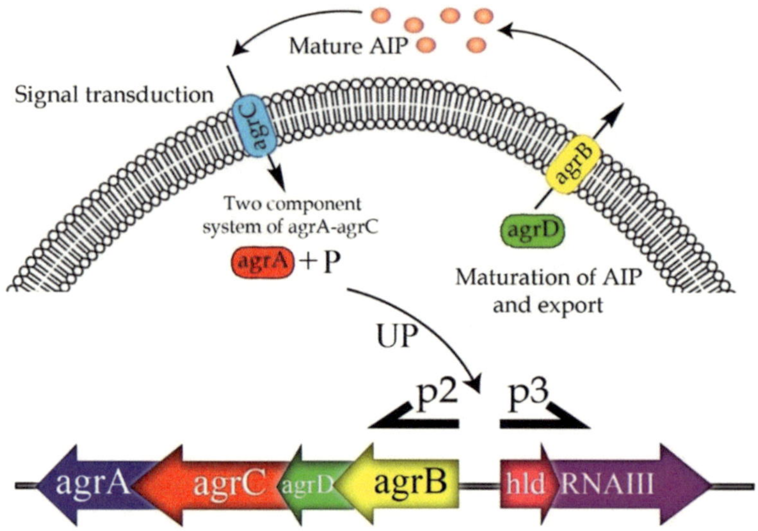

Figure 2.
agr quorum sensing system in Staphylococcus aureus. Autoinducer peptides (AIPs) are produced from the agrD precursor by agrB. Mature AIPs are then exported outside the cell till their concentration reaches a threshold when the two-component system (agrC/agrA) becomes activated. Afterwards, agrA is phosphorylated, enabling it to activate transcription of the P2 and P3 promotors (upregulation). Also, agrA is involved in encoding phenol-soluble modulin peptides (via increasing transcription of psmα and psmβ operons). RNAIII is responsible for regulation of most agr targets as well as delta-toxin (hld).

accessory gene regulator (*agr*) and *sae* are capable of sensing environmental stimuli, whereas *arl*RS is thought to play a part in antibiotic resistance as well as autolytic activity. The last two-component system, *srr*AB, has a role in energy metabolism and RNAIII inhibition [45]. In addition, *agr* is responsible for controlling virulence factor gene expression by *S. aureus*. The *agr* locus is a density-dependent system, composed of two QS components [46]. There are four main subgroups of *agr*, each of which produces a distinctive AIP. Meanwhile, a two-component system comprising *agr*A and *agr*C is responsible for AIP identification. These AIPs are similar in terms of thiolactone ring structure but differ in amino acid sequence. Moreover, QS is regulated via the *agr* locus, which comprises two transcripts, RNAII and RNAIII. Their expression is induced via P2 and P3 promotors, respectively [45, 47–51]. Activation of *agr* is not limited to AIPs, as additional proteins such as SarA, SrrAB, and other environmental factors can also activate the system [52, 53]. Initially, when *S. aureus* population density is not sufficiently high to induce *agr* expression, colonization occurs through the production of surface proteins, followed by *agr* expression upon increasing cell density. Therefore, *agr* timing adaptation is an indicative of infection progression [54, 55].

In the *agr* system, RNAIII has an important function as an intercellular effector in controlling target gene expression. It also controls other virulence factors, including protein A, Rot protein, leukocidins, enterotoxins, and alpha toxin [50]. Moreover, the four genes *agr*B, *agr*C, *agr*D, and *agr*A are located in the RNAII operon. Initially, *agr*D encodes a 46-amino acid peptide (pro-AIP), which is later processed to yield a 9-amino acid residue. This AIP precursor undergoes modification to C-terminal cleavage before exportation from the cell via *agr*B (a transmembrane endopeptidase). AIP signaling molecules are released into the extracellular environment and have accumulated there until a threshold concentration is reached when they are detected by specific sensors. *agr*C, a transmembrane histidine kinase protein, is phosphorylated and attaches to AIPs, thereby enabling gene regulation in QS to be followed. This is also responsible for the activation of *agr*A, which is a response regulator [8, 49, 50, 56, 57]. In an autofeedback cycle, upregulation of RNAII and RNAIII transcription is driven by the binding of *agr*A to P2 and P3 promoters, respectively [50]. In short, the simultaneous activation of *agr*A and *agr*C, which act as transcription factors for RNAIII, induces the RNAII operon [49, 58]. Upon activation, RNAIII can trigger the production of alpha toxin. Meanwhile, the RNAIII is able to quench the expression of certain surface virulence factors (including coagulase, peptide A and FNPA, B; [59].

A further activation pathway has been reported in various Gram-positive bacteria. This involves interaction between signaling molecules and receptors inside the cell, after which the expressed products are transported to the external environment [60, 61]. This is exemplified by *Enterococcus faecalis*, in which the interaction between peptides and PrgX proteins alters the activity of conjugative plasmids [62, 63] and by Phr peptides acting as phosphatase inhibitors in *Bacillus* species [62, 63]. Finally, a strong relationship between *agr* and σ^B, a biofilm formation regulator, has been identified [64]. Formation and dispersal of biofilm are associated with the downregulation and upregulation of *agr*, respectively [65, 66].

1.3 Autoinducer-2 in Gram-positive and Gram-negative bacteria

AI-2 is found in both Gram-positive and Gram-negative bacteria, where it facilitates intra- and inter-species communication [67, 68]. AI-2 signals have been described as providing an "interconversion nature", meaning that this molecule is utilized by different bacteria as a universal tool for communication [68]. Support for this notion comes from the observation that, unlike for single-species oral biofilm formation, in mixed populations of *Porphyromonas gingivalis* and *Streptococcus gordonii*, LuxS expression by each species is required. Further evidence shows that if there is a deficiency of *Streptococcus mutans,* other species of oral bacteria supplement with *luxS* mutation in biofilm formation [69].

In this system, the enzyme LuxS catalyzes the synthesis of AI-2 or its precursor 4,5-dihydroxy-2,3-pentanedione [70]. Two receptors, LuxP (a periplasmic-binding protein) and LsrB, are detected. Biofilm formation, virulence factor production, and other density-dependent phenotypes are attributed to the former, with delivery of AI-2 into cells ascribed to the latter [67, 70, 71]. They differ in structure, exemplified by LuxP-AI-2 in *Vibrio harveyi* being composed of furanosyl borate diester, whereas LsrB-AI-2 in *Salmonella typhimurium* lacks boron [71, 72]. Molecular analysis indicates that the type of AI-2 varies with bacterial species [72]. Multiple bacteria have been identified that can react to AI-2, including *Staphylococcus epidermidis*, *Helicobacter pylori*, *Bacillus subtilis, Pseudomonas aeruginosa*, *Campylobacter jejuni*, *S. mutans,* and *Listeria monocytogenes* [73–79]. To date, most information on this system comes from *V. harveyi* [69], for which three-channel quorum sensing is proposed, involving AI-1, AI-2 and, cholerae AI-1 [80].

The *V. harveyi* protein LuxQ has a cytoplasmic histidine-kinase domain, a response regulatory domain, and a periplasmic sensor domain. Interestingly, upon binding to the AI-2, LuxQ functions as a kinase and as a phosphatase at low and high cell densities, respectively [70, 81]. Another protein known as LuxP is able to modify LuxQ activity (through a histidine-kinase sensor), and it is this union that regulates

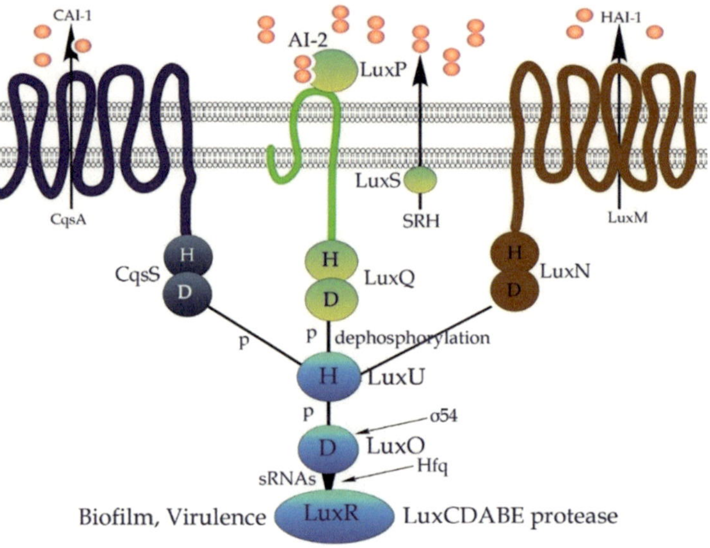

Figure 3.
AI-2 quorum sensing system in Vibrio harveyi. The three autoinducers HAI-1, AI-2 and CAI-1 are synthesized by LuxM, LuxS and CqsA, respectively. LuxS produces AI-2 by converting S-ribosylhomocysteine (SRH) to dihydroxypentane-2,3-dione (DPD) in the cell cytoplasm. This occurs when LuxS participates in the activated methyl cycle, which generates and recycles methyl donors. DPD is a by-product of the LuxS reaction that produces SRH. Later, DPD undergoes cyclization and rearranges without enzymatic support to produce AI-2 prior to export across the outer membrane. A two-component signal regulator is responsible for the responding pathway in vibrio spp., while for Salmonella enterica this is identified as an ABC transporter. In V. harveyi, furanosyl-borat-diester and periplasmic LuxP together form active AI-2, inducing phosphatase activity in LuxQ. This leads to phosphate transfer from LuxU to LuxO, which is the response regulator. Finally, several cell changes take place, including bioluminescence. In contrast, when cell density is low and there is no AI-2, phosphorylated LuxO as well as σ^{54} produce small regulatory RNAs. Their interaction with $LuxR_{Vh}$ mRNA causes destabilization of Hfq-dependent chaperone proteins. This results in suppression of transcription of the lux operon and a reduction in bioluminescence. Meanwhile, dephosphorylated LuxO, the level of which increases in the presence of AI-2, reverses the flow of phosphate.

the AI-2 QS regulon (**Figure 3**) [70, 82]. Following the conversion of LuxQ activity from kinase to phosphatase via the transmembrane sensor histidine kinase, LuxP bound to AI-2 regulates gene expression of phenotypes such as biofilm formation and bioluminescence [67, 70, 83].

2. QS and biofilm

Multiple factors benefit bacterial colonies that adopt a multicellular lifestyle rather than remain planktonic. Bacterial cells embedded within biofilm are protected from detrimental factors, whereas nutrient-deficient conditions and hostile environments are both noted among driver factors for biofilm production [84]. A crucial component of mature *S. aureus* biofilm is the extracellular matrix. This is composed of eDNA, polysaccharide intercellular adhesin, and other proteins. It is the most stable, thus, a problematic stage to treat. To reduce biofilm mass, detachment follows, in which QS as well as nuclease and protease enzymes play a significant part [85–88]. The cell-to-cell signaling of QS pertains to all biofilm formation stages. A key role in the initiation is the communication between bacteria through the detection of AIPs [89]. Chronic infection of *S. aureus* as well as biofilm formation is linked with low activity of *agr* QS [50]. *In vivo* studies have demonstrated the importance of the *agr* system to disease progression. Although upregulation by *agr* has a role in acute infections, downregulation is involved in biofilm formation [66, 90, 91]. In the dispersal stage, which is directly under-regulation of *agr*, isolation of new cells is ascribed to P3 promoters via the production of proteases and glucose depletion [92]. Hence, there is a direct relation between QS activation and the transition between biofilm and planktonic cell lifestyles. Thus, it remains to be determined whether QS quenching results in biofilm blockage [93, 94].

3. Anti-QS approaches

Because of widespread heightened antimicrobial resistance, the conventional means of treating bacterial infection, antibiotic therapy, is now increasingly impractical, such that alternative approaches are being considered [95]. The presence of biofilm, efflux pumps, and persister cells each exacerbate drug resistance [96]. Targeting QS by disturbing cell-cell communication is a way to combat biofilm [97]. Moreover, the effectiveness of different potential inhibitors against QS has been reported [98]. Various strategies are proposed to disrupt QS, including receptor inactivation, signal inhibition (by natural or synthetic inhibitors), signal degradation by quorum quenching enzymes, blocking QS by antibodies, and applying antibiotics as a cotreatment [98, 99].

Targeting AIPs is a good way of treating QS and considerable effort has been made to date to find inhibitors [100]. A known approach suggested in this context is to cope with RNAIII, due to its key role in QS. Reportedly, RNAIII inhibitory peptides (RIPs) have shown inhibitory effects on *agr* and biofilm. It is believed that targeting this molecule will diminish the production of some virulence factors and toxins [101, 102]. Similarly, based on the inhibition of *agr* of another subgroup [103], natural and synthetic AIPs may be introduced as potential inhibitors. Different inhibitors include nonpeptidic (P3 inhibitor) and synthetic molecules, cyclic dipeptides (from *Lactobacillus*), ambuic acid (a fungal extract), licochalcone A (LicA, a plant extract), antivirulence agents such as naphthalene and biaryl compounds, organic compounds (by interfering with *agr*-DNA binding), and savirin (*S. aureus* virulence inhibitor). Each of these actively inhibits QS in *S. aureus* [104–110]. Monoclonal antibodies, applied as both passive and active immunotherapeutic regimens, have also yielded promising results [49]. Collectively, many approaches that target *agr* and AIPs have been tested. Targeting AIPs via extracellular therapy is advantageous over targeting *agr*, as complications of intercellular therapy such as degradation do not arise.

Another treatment approach for Gram-negative bacteria is based on phenolic compounds. When tested extensively against AHL QS in *P. aeruginosa*, the novel phenolic derivative GM-50 reduces biofilm-related virulence, thereby enhancing antibiotic efficacy [111]. In addition, food-associated bacteria such as lactobacilli can exploit antibiofilm activity by interfering with AHL QS [112]. The efficacy of probiotics against QS has been indicated in previous studies. Presumably, they exert their effects via secretion of metabolites and microencapsulation [113, 114].

Targeting AI-2 lessens the pathogenicity of different bacterial species [115]. Various natural products such as D-galactose and furanocoumarin (reducing AI-2 synthesis), apigenin, hexadecenoic acid, and citral have shown promise at inhibiting *V. harveyi* QS [116–120]. In terms of chemicals, halogenated furanones are effective against AHL and AI-2, subsequently affecting biofilm formation [121]. In *C. jejuni*, two fatty acids, decanoic acid and lauric acid, were found to be useful against AI-2 at 100 ppm (preventing 90% of AI-2 activity). As a result, biofilm formation and motility of the bacterium were reduced substantially [115]. Similarly, different naturally occurring compounds including monoterpenoid glycosides, emodin, and antimicrobial peptides showed satisfactory inhibition of LuxS/AI-2 in *Streptococcus suis* [122–125].

Currently, there is no drug approved for clinical use, although research and development efforts are continuously making progress toward this goal. As a consequence of administering anti-QS drugs, bacterial virulence (selective pressure will result in no further negative implications) applied should decrease, which is of great importance when seeking novel, effective treatments [111, 126].

4. Conclusions

The complex adaptive regulatory system of QS stands out as the most pivotal mechanism of pathogenicity exhibited by bacteria [127]. Regarding therapy, because of the emergence and widespread prevalence of antibiotic resistance, cotreatment with alternatives as well as surgical removal of infected tissue surrounding implanted medical devices, is being increasingly used. Quenching and inhibitory substances suppress the virulence and pathogenicity of those bacterial pathogens that use QS. Because QS has a critical role in many physiological behaviors such as biofilm formation, exoenzyme secretion, siderophore functioning, membrane vesicle formation, swarming, and sporulation, QQ is becoming a popular strategy [128]. Thus, an in-depth knowledge of biofilm, sensitive antibiotics, penetration, and anti-QS agents will help to inform antimicrobial therapies to overcome biofilm infection [129].

Multiple activities of anti-QS agents have been identified, for instance, QS receptor inactivation, QS signal inhibition, degradation of QS signals, and antibodies to block QS, as well as combination therapies such as flavonoids or immucillin A in *P. aeruginosa,* lactonase in *Acinetobacter baumanni*, AP4-24H11 in *S. aureus,* and farensol with ß-lactamase antibiotics in *S. aureus* [130–134]. Given this premise, a QS inhibitor can modulate gene regulation via either of two strategies: interposition with signal generation and signal reception [135, 136]. Notably, many QS inhibitors, such as furanones and halogenated and acylated furan structures, are improved by competing with the AHL pheromone in *P. aeruginosa* [137, 138]. Furthermore, RIPs have shown promise against *S. aureus* [139].

Negative aspects of disturbing the QS system should be considered. Inadvertent or unregulated modulation of microbiota through the use of QS quenching compounds or inhibitors may cause a disequilibrium of normal microflora. This concept developed as AI-2 molecules resemble bacterial presence to provide microflora [128, 140]. At the same time, pathogenicity tends to increase by applying quenching agents that may contribute to the long-term survival of *S. aureus* [141–144]. In particular, staphylococcal QS *agr* mutant strains tend to develop persister forms as well as raised biofilm production [143, 145]. A possible strategy is to apply QS quenching only in the absence of biofilm. This stems from the observation of applying selective pressure to preserve *agr* as a planktonic form rather than in biofilm [146].

An important clinical consideration is to determine the strain susceptibility and optimal form of treatment, otherwise, the patient's condition may worsen [102]. In addition, limitations and challenges should be carefully weighed. For example, in

S. aureus, the type of condition should be considered, as *agr* performs contrary roles in biofilm and chronic infection. Of note, most studies on QS drugs have been carried out using a single laboratory strain. Although such research models provide valuable information, it is challenging to extrapolate with confidence to clinical settings in, for instance, the case of AIPs in *S. aureus,* as species subgroups are identified. Finally, the selectivity and safety of QS inhibitors, while minimizing disturbance of microflora, are important factors for human usage. Designing a library of QS inhibitors and determining their IC_{50} values is a suggested area for future research.

Conflict of interest

The authors declare no conflict of interest.

Appendices and Nomenclature

AI	Autoinducer.
Agr	Accessory gene regulator
AgrA	Response regulator
AgrB	Membrane-associated export protein, processes AgrD into AIP
AgrC	Membrane-bound histidine kinase receptor
AgrD	Propeptide gene for AIP
AHL	Acyl-homoserine lactone
AIP	Auto-inducing peptide
PIA	Polysaccharide intercellular adhesin
QS	Quorum-sensing
AI-2	Autoinducer-2
RIP	RNAIII inhibitory peptide

Author details

Zahra Sedarat[1]* and Andrew W. Taylor-Robinson[2,3,4]

1 Cellular and Molecular Research Centre, Shahrekord University of Medical Sciences, Shahrekord, Iran

2 College of Health Sciences, VinUniversity, Hanoi, Vietnam

3 Center for Global Health, University of Pennsylvania, Philadelphia, USA

4 College of Health and Human Sciences, Charles Darwin University, Casuarina, Australia

*Address all correspondence to: sedaratzahra@gmail.com

References

[1] Tomasz A. Control of the competent state in *Pneumococcus* by a hormone-like cell product: An example for a new type of regulatory mechanism in bacteria. Nature. 1965;**208**(5006):155-159

[2] Dunny GM. The peptide pheromone-inducible conjugation system of *Enterococcus faecalis* plasmid pCF10: Cell-cell signalling, gene transfer, complexity and evolution. Philosophical Transactions of the Royal Society of London. Series B, Biological Sciences. 2007;**362**(1483):1185-1193

[3] Nealson KH. Autoinduction of bacterial luciferase. Occurrence, mechanism and significance. Archives of Microbiology. 1977;**112**(1):73-79

[4] Rutherford ST, Bassler BL. Bacterial quorum sensing: Its role in virulence and possibilities for its control. Cold Spring Harbor Perspectives in Medicine. 2012;**2**(11):a012427

[5] Novick RP, Geisinger E. Quorum sensing in staphylococci. Annual Review of Genetics. 2008;**42**:541-564

[6] Ng WL, Bassler BL. Bacterial quorum-sensing network architectures. Annual Review of Genetics. 2009;**43**:197-222

[7] Williams P, Cámara M. Quorum sensing and environmental adaptation in *Pseudomonas aeruginosa*: A tale of regulatory networks and multifunctional signal molecules. Current Opinion in Microbiology. 2009;**12**(2):182-191

[8] Miller MB, Bassler BL. Quorum sensing in bacteria. Annual Review of Microbiology. 2001;**55**:165-199

[9] Arevalo-Ferro C, Hentzer M, Reil G, Görg A, Kjelleberg S, Givskov M, et al. Identification of quorum-sensing regulated proteins in the opportunistic pathogen *Pseudomonas aeruginosa* by proteomics. Environmental Microbiology. 2003;**5**(12):1350-1369

[10] Striednig B, Hilbi H. Bacterial quorum sensing and phenotypic heterogeneity: How the collective shapes the individual. Trends in Microbiology. 2022;**30**(4):379-389

[11] Lixa C, Mujo A, Anobom CD, Pinheiro AS. A structural perspective on the mechanisms of quorum sensing activation in bacteria. Anais da Academia Brasileira de Ciências. 2015;**87**(4):2189-2203

[12] Subramani R, Jayaprakashvel M. Bacterial quorum sensing: Biofilm formation, survival behaviour and antibiotic resistance. Implication of Quorum Sensing and Biofilm Formation in Medicine, Agriculture and Food Industry. Singapore: Springer; 2019;**2019**:21-37

[13] Xavier KB, Bassler BL. LuxS quorum sensing: More than just a numbers game. Current Opinion in Microbiology 2003;**6**(2):191-197

[14] Papenfort K, Bassler BL. Quorum sensing signal-response systems in Gram-negative bacteria. Nature Reviews Microbiology. 2016;**14**(9):576-588

[15] Girard L, Blanchet E, Stien D, Baudart J, Suzuki M, Lami R. Evidence of a large diversity of N-acyl-homoserine lactones in symbiotic *Vibrio fischeri* strains associated with the squid *Euprymna scolopes*. Microbes and Environments. 2019;**34**(1):99-103

[16] Papenfort K, Vogel J. Regulatory RNA in bacterial pathogens. Cell Host & Microbe. 2010;**8**(1):116-127

[17] Case RJ, Labbate M, Kjelleberg S. AHL-driven quorum-sensing circuits: Their frequency and function among the Proteobacteria. The ISME Journal. 2008;**2**(4):345-349

[18] Callahan SM, Dunlap PV. LuxR- and acyl-homoserine-lactone-controlled non-lux genes define a quorum-sensing regulon in *Vibrio fischeri*. Journal of Bacteriology. 2000;**182**(10):2811-2822

[19] Lupp C, Urbanowski M, Greenberg EP, Ruby EG. The *Vibrio fischeri* quorum-sensing systems Ain and lux sequentially induce luminescence gene expression and are important for persistence in the squid host. Molecular Microbiology. 2003;**50**(1):319-331

[20] Chen G, Swem LR, Swem DL, Stauff DL, O'Loughlin CT, Jeffrey PD, et al. A strategy for antagonizing quorum sensing. Molecular Cell. 2011;**42**(2):199-209

[21] McInnis CE, Blackwell HE. Thiolactone modulators of quorum sensing revealed through library design and screening. Bioorganic & Medicinal Chemistry. 2011;**19**(16):4820-4828

[22] Fuqua WC, Winans SC, Greenberg EP. Quorum sensing in bacteria: The LuxR-LuxI family of cell density-responsive transcriptional regulators. Journal of Bacteriology. 1994;**176**(2):269-275

[23] Miyashiro T, Ruby EG. Shedding light on bioluminescence regulation in *Vibrio fischeri*. Molecular Microbiology. 2012;**84**(5):795-806

[24] Sitnikov DM, Schineller JB, Baldwin TO. Transcriptional regulation of bioluminesence genes from *Vibrio fischeri*. Molecular Microbiology. 1995;**17**(5):801-812

[25] Gould TA, Schweizer HP, Churchill ME. Structure of the *Pseudomonas aeruginosa* acyl-homoserinelactone synthase LasI. Molecular Microbiology. 2004;**53**(4):1135-1146

[26] Watson WT, Minogue TD, Val DL, von Bodman SB, Churchill ME. Structural basis and specificity of acyl-homoserine lactone signal production in bacterial quorum sensing. Molecular Cell. 2002;**9**(3):685-694

[27] Fuqua C, Winans SC, Greenberg EP. Census and consensus in bacterial ecosystems: The LuxR-LuxI family of quorum-sensing transcriptional regulators. Annual Review of Microbiology. 1996;**50**:727-751

[28] Schaefer AL, Val DL, Hanzelka BL, Cronan JE Jr, Greenberg EP. Generation of cell-to-cell signals in quorum sensing: Acyl homoserine lactone synthase activity of a purified *Vibrio fischeri* LuxI protein. Proceedings of the National Academy of Sciences of the United States of America. 1996;**93**(18):9505-9509

[29] Urbanowski ML, Lostroh CP, Greenberg EP. Reversible acyl-homoserine lactone binding to purified *Vibrio fischeri* LuxR protein. Journal of Bacteriology. 2004;**186**(3):631-637

[30] Choi SH, Greenberg EP. The C-terminal region of the *Vibrio fischeri* LuxR protein contains an inducer-independent lux gene activating domain. Proceedings of the National Academy of Sciences of the United States of America. 1991;**88**(24):11115-11119

[31] Finney AH, Blick RJ, Murakami K, Ishihama A, Stevens AM. Role of the C-terminal domain of the alpha subunit of RNA polymerase in LuxR-dependent transcriptional activation

of the lux operon during quorum sensing. Journal of Bacteriology. 2002;**184**(16):4520-4528

[32] Hanzelka BL, Greenberg EP. Evidence that the N-terminal region of the *Vibrio fischeri* LuxR protein constitutes an autoinducer-binding domain. Journal of Bacteriology. 1995;**177**(3):815-817

[33] Nijvipakul S, Wongratana J, Suadee C, Entsch B, Ballou DP, Chaiyen P. LuxG is a functioning flavin reductase for bacterial luminescence. Journal of Bacteriology. 2008;**190**(5):1531-1538

[34] Tinikul R, Chunthaboon P, Phonbuppha J, Paladkong T. Bacterial luciferase: Molecular mechanisms and applications. The Enzymes. 2020;**47**:427-455

[35] Pérez PD, Hagen SJ. Heterogeneous response to a quorum-sensing signal in the luminescence of individual *Vibrio fischeri*. PLoS One. 2010;**5**(11):e15473

[36] Boylan M, Miyamoto C, Wall L, Graham A, Meighen E, Lux C. D and E genes of the *Vibrio fischeri* luminescence operon code for the reductase, transferase, and synthetase enzymes involved in aldehyde biosynthesis. Photochemistry and Photobiology. 1989;**49**(5):681-688

[37] Lawan N, Tinikul R, Surawatanawong P, Mulholland AJ, Chaiyen P. QM/MM molecular modeling reveals mechanism insights into flavin peroxide formation in bacterial luciferase. Journal of Chemical Information and Modeling. 2022;**62**(2):399-411

[38] Egland KA, Greenberg EP. Quorum sensing in *Vibrio fischeri*: Elements of the luxl promoter. Molecular Microbiology. 1999;**31**(4):1197-1204

[39] Baldwin TO, Devine JH, Heckel RC, Lin JW, Shadel GS. The complete nucleotide sequence of the lux regulon of *Vibrio fischeri* and the luxABN region of *Photobacterium leiognathi* and the mechanism of control of bacterial bioluminescence. Journal of Bioluminescence and Chemiluminescence. 1989;**4**(1):326-341

[40] Whitehead NA, Barnard AML, Slater H, Simpson NJL, Salmond GPC. Quorum-sensing in Gram-negative bacteria. FEMS Microbiology Reviews. 2001;**25**(4):365-404

[41] Ji G, Beavis RC, Novick RP. Cell density control of staphylococcal virulence mediated by an octapeptide pheromone. Proceedings of the National Academy of Sciences of the United States of America. 1995;**92**(26):12055-12059

[42] Okada M, Sato I, Cho SJ, Iwata H, Nishio T, Dubnau D, et al. Structure of the *Bacillus subtilis* quorum-sensing peptide pheromone ComX. Nature Chemical Biology. 2005;**1**(1):23-24

[43] Bouillaut L, Perchat S, Arold S, Zorrilla S, Slamti L, Henry C, et al. Molecular basis for group-specific activation of the virulence regulator PlcR by PapR heptapeptides. Nucleic Acids Research. 2008;**36**(11):3791-3801

[44] Prazdnova EV, Gorovtsov AV, Vasilchenko NG, Kulikov MP, Statsenko VN, Bogdanova AA, et al. Quorum-sensing inhibition by gram-positive bacteria. Microorganisms. 2022;**10**(2):350

[45] Yarwood JM, Schlievert PM. Quorum sensing in *Staphylococcus* infections. The Journal of Clinical Investigation. 2003;**112**(11):1620-1625

[46] Bibalan MH, Shakeri F, Javid N, Ghaemi A, Ghaemi EA. Accessory gene

regulator types of *Staphylococcus aureus* isolated in Gorgan, North of Iran. Journal of Clinical and Diagnostic Research : JCDR. 2014;**8**(4):Dc07-Dc09

[47] Håvarstein LS, Coomaraswamy G, Morrison DA. An unmodified heptadecapeptide pheromone induces competence for genetic transformation in *Streptococcus pneumoniae*. Proceedings of the National Academy of Sciences of the United States of America. 1995;**92**(24):11140-11144

[48] Thoendel M, Horswill AR. Biosynthesis of peptide signals in gram-positive bacteria. Advances in Applied Microbiology. 2010;**71**:91-112

[49] Kirchdoerfer RN, Garner AL, Flack CE, Mee JM, Horswill AR, Janda KD, et al. Structural basis for ligand recognition and discrimination of a quorum-quenching antibody. The Journal of Biological Chemistry. 2011;**286**(19):17351-17358

[50] Le KY, Otto M. Quorum-sensing regulation in staphylococci: An overview. Frontiers in Microbiology. 2015;**6**:1174

[51] Ji G, Beavis R, Novick RP. Bacterial interference caused by autoinducing peptide variants. Science (New York, N.Y.). 1997;**276**(5321):2027-2030

[52] Vijayakumar K, Muhilvannan S, Arun VM. Hesperidin inhibits biofilm formation, virulence and staphyloxanthin synthesis in methicillin resistant *Staphylococcus aureus* by targeting SarA and CrtM: An in vitro and in silico approach. World Journal of Microbiology & Biotechnology. 2022;**38**(3):44

[53] Yarwood JM, McCormick JK, Schlievert PM. Identification of a novel two-component regulatory system that acts in global regulation of virulence factors of *Staphylococcus aureus*. Journal of Bacteriology. 2001;**183**(4):1113-1123

[54] Cheung GY, Kretschmer D, Duong AC, Yeh AJ, Ho TV, Chen Y, et al. Production of an attenuated phenol-soluble modulin variant unique to the MRSA clonal complex 30 increases severity of bloodstream infection. PLoS Pathogens. 2014;**10**(8):e1004298

[55] Fowler VG Jr, Sakoulas G, McIntyre LM, Meka VG, Arbeit RD, Cabell CH, et al. Persistent bacteremia due to methicillin-resistant *Staphylococcus aureus* infection is associated with agr dysfunction and low-level in vitro resistance to thrombin-induced platelet microbicidal protein. The Journal of Infectious Diseases. 2004;**190**(6):1140-1149

[56] Saenz HL, Augsburger V, Vuong C, Jack RW, Götz F, Otto M. Inducible expression and cellular location of AgrB, a protein involved in the maturation of the staphylococcal quorum-sensing pheromone. Archives of Microbiology. 2000;**174**(6):452-455

[57] Zhang L, Ji G. Identification of a staphylococcal AgrB segment(s) responsible for group-specific processing of AgrD by gene swapping. Journal of Bacteriology. 2004;**186**(20):6706-6713

[58] Koenig RL, Ray JL, Maleki SJ, Smeltzer MS, Hurlburt BK. *Staphylococcus aureus* AgrA binding to the RNAIII-agr regulatory region. Journal of Bacteriology. 2004;**186**(22):7549-7555

[59] Kutty SK, Barraud N, Pham A, Iskander G, Rice SA, Black DS, et al. Design, synthesis, and evaluation of fimbrolide-nitric oxide donor hybrids as antimicrobial agents. Journal of Medicinal Chemistry. 2013;**56**(23):9517-9529

[60] Pottathil M, Lazazzera BA. The extracellular Phr peptide-Rap phosphatase signaling circuit of *Bacillus subtilis*. Frontiers in Bioscience: A Journal and Virtual Library. 2003;**8**:d32-d45

[61] Perego M. Forty years in the making: Understanding the molecular mechanism of peptide regulation in bacterial development. PLoS Biology. 2013;**11**(3):e1001516

[62] Slamti L, Perchat S, Huillet E, Lereclus D. Quorum sensing in *Bacillus thuringiensis* is required for completion of a full infectious cycle in the insect. Toxins. 2014;**6**(8):2239-2255

[63] Dunny GM. Enterococcal sex pheromones: Signaling, social behavior, and evolution. Annual Review of Genetics. 2013;**47**:457-482

[64] Shaw LN, Aish J, Davenport JE, Brown MC, Lithgow JK, Simmonite K, et al. Investigations into sigmaB-modulated regulatory pathways governing extracellular virulence determinant production in *Staphylococcus aureus*. Journal of Bacteriology. 2006;**188**(17):6070-6080

[65] Lee HS, Song HS, Lee HJ, Kim SH, Suh MJ, Cho JY, et al. Comparative study of the difference in behavior of the accessory gene regulator (Agr) in USA300 and USA400 community-associated methicillin-resistant *Staphylococcus aureus* (CA-MRSA). Journal of Microbiology and Biotechnology. 2021;**31**(8):1060-1068

[66] Periasamy S, Joo HS, Duong AC, Bach TH, Tan VY, Chatterjee SS, et al. How *Staphylococcus aureus* biofilms develop their characteristic structure. Proceedings of the National Academy of Sciences of the United States of America. 2012;**109**(4):1281-1286

[67] Pereira CS, Thompson JA, Xavier KB. AI-2-mediated signalling in bacteria. FEMS Microbiology Reviews. 2013;**37**(2):156-181

[68] Xavier KB, Bassler BL. Interference with AI-2-mediated bacterial cell-cell communication. Nature. 2005;**437**(7059):750-753

[69] Horinouchi S, Ueda K, Nakayama J, Ikeda T. 4.07 - cell-to-cell communications among microorganisms. In: Liu H-W, Mander L, editors. Comprehensive Natural Products II. Oxford: Elsevier; 2010. pp. 283-337

[70] Neiditch MB, Federle MJ, Miller ST, Bassler BL, Hughson FM. Regulation of LuxPQ receptor activity by the quorum-sensing signal autoinducer-2. Molecular Cell. 2005;**18**(5):507-518

[71] Pereira CS, de Regt AK, Brito PH, Miller ST, Xavier KB. Identification of functional LsrB-like autoinducer-2 receptors. Journal of Bacteriology. 2009;**191**(22):6975-6987

[72] Chen X, Schauder S, Potier N, Van Dorsselaer A, Pelczer I, Bassler BL, et al. Structural identification of a bacterial quorum-sensing signal containing boron. Nature. 2002;**415**(6871):545-549

[73] Plummer PJ. LuxS and quorum-sensing in *Campylobacter*. Frontiers in Cellular and Infection Microbiology. 2012;**2**:22

[74] Anderson JK, Huang JY, Wreden C, Sweeney EG, Goers J, Remington SJ, et al. Chemorepulsion from the quorum signal autoinducer-2 promotes *Helicobacter pylori* biofilm dispersal. MBio. 2015;**6**(4):e00379

[75] Yuan K, Hou L, Jin Q, Niu C, Mao M, Wang R, et al. Comparative transcriptomics analysis of *Streptococcus*

mutans with disruption of LuxS/AI-2 quorum sensing and recovery of methyl cycle. Archives of Oral Biology. 2021;**127**:105137

[76] Li H, Li X, Song C, Zhang Y, Wang Z, Liu Z, et al. Autoinducer-2 facilitates *Pseudomonas aeruginosa* PAO1 pathogenicity in vitro and in vivo. Frontiers in Microbiology. 2017;**8**:1944

[77] Lombardía E, Rovetto AJ, Arabolaza AL, Grau RR. A LuxS-dependent cell-to-cell language regulates social behavior and development in *Bacillus subtilis*. Journal of Bacteriology. 2006;**188**(12):4442-4452

[78] Li M, Villaruz AE, Vadyvaloo V, Sturdevant DE, Otto M. AI-2-dependent gene regulation in *Staphylococcus epidermidis*. BMC Microbiology. 2008;**8**:4

[79] Garmyn D, Gal L, Lemaitre JP, Hartmann A, Piveteau P. Communication and autoinduction in the species *Listeria* monocytogenes: A central role for the agr system. Communicative & Integrative Biology. 2009;**2**(4):371-374

[80] Defoirdt T, Boon N, Sorgeloos P, Verstraete W, Bossier P. Quorum sensing and quorum quenching in *Vibrio harveyi*: Lessons learned from in vivo work. The ISME Journal. 2008;**2**(1):19-26

[81] Bassler BL, Wright M, Silverman MR. Multiple signalling systems controlling expression of luminescence in *Vibrio harveyi*: Sequence and function of genes encoding a second sensory pathway. Molecular Microbiology. 1994;**13**(2):273-286

[82] Neiditch MB, Federle MJ, Pompeani AJ, Kelly RC, Swem DL, Jeffrey PD, et al. Ligand-induced asymmetry in histidine sensor kinase complex regulates quorum sensing. Cell. 2006;**126**(6):1095-1108

[83] Miranda V, Torcato IM, Xavier KB, Ventura MR. Synthesis of d-desthiobiotin-AI-2 as a novel chemical probe for autoinducer-2 quorum sensing receptors. Bioorganic Chemistry. 2019;**92**:103200

[84] Paluch E, Rewak-Soroczyńska J, Jędrusik I, Mazurkiewicz E, Jermakow K. Prevention of biofilm formation by quorum quenching. Applied Microbiology and Biotechnology. 2020;**104**(5):1871-1881

[85] Lauderdale KJ, Boles BR, Cheung AL, Horswill AR. Interconnections between sigma B, agr, and proteolytic activity in *Staphylococcus aureus* biofilm maturation. Infection and Immunity. 2009;**77**(4):1623-1635

[86] Hooshdar P, Kermanshahi R, Ghadam P, Khosravi K. A review on production of exopolysaccharide and biofilm in probiotics like *Lactobacilli and* methods of analysis. Biointerface Research in Applied Chemistry. 2020;**2020**:10

[87] Sedarat Z, Taylor-Robinson AW. Biofilm formation by pathogenic bacteria: Applying a *Staphylococcus aureus* model to appraise potential targets for therapeutic intervention. Pathogens. 2022;**11**(4):388

[88] Joo HS, Fu CI, Otto M. Bacterial strategies of resistance to antimicrobial peptides. Philosophical Transactions of the Royal Society of London. Series B, Biological Sciences. 2016;**371**:1695

[89] Kong KF, Vuong C, Otto M. *Staphylococcus* quorum sensing in biofilm formation and infection. International Journal of Medical Microbiology. 2006;**296**(2-3):133-139

[90] Heyer G, Saba S, Adamo R, Rush W, Soong G, Cheung A, et al.

Staphylococcus aureus agr and sarA functions are required for invasive infection but not inflammatory responses in the lung. Infection and Immunity. 2002;**70**(1):127-133

[91] Montgomery CP, Boyle-Vavra S, Daum RS. Importance of the global regulators Agr and SaeRS in the pathogenesis of CA-MRSA USA300 infection. PLoS One. 2010;**5**(12):e15177

[92] Thoendel M, Kavanaugh JS, Flack CE, Horswill AR. Peptide signaling in the staphylococci. Chemical Reviews. 2011;**111**(1):117-151

[93] Boles BR, Horswill AR. Agr-mediated dispersal of *Staphylococcus aureus* biofilms. PLoS Pathogens. 2008;**4**(4):e1000052

[94] Boles BR, Horswill AR. Staphylococcal biofilm disassembly. Trends in Microbiology. 2011;**19**(9):449-455

[95] Fan Q, Zuo J, Wang H, Grenier D, Yi L, Wang Y. Contribution of quorum sensing to virulence and antibiotic resistance in zoonotic bacteria. Biotechnology Advances. 2022;**59**:107965

[96] Akinbobola AB, Sherry L, McKay WG, Ramage G, Williams C. Tolerance of *Pseudomonas aeruginosa* in in-vitro biofilms to high-level peracetic acid disinfection. The Journal of Hospital Infection. 2017;**97**(2):162-168

[97] Yada S, Kamalesh B, Sonwane S, Guptha I, Swetha RK. Quorum sensing inhibition, relevance to periodontics. Journal of International Oral Health : JIOH. 2015;**7**(1):67-69

[98] Jiang Q, Chen J, Yang C, Yin Y, Yao K. Quorum sensing: A prospective therapeutic target for bacterial diseases. BioMed Research International. 2019;**2019**:2015978

[99] Escobar-Muciño E, Arenas-Hernández MMP, Luna-Guevara ML. Mechanisms of inhibition of quorum sensing as an alternative for the control of *E. coli* and *Salmonella*. Microorganisms. 2022;**10**(5):884

[100] Milly TA, Tal-Gan Y. Targeting peptide-based quorum sensing systems for the treatment of gram-positive bacterial infections. Peptide Science. 2023;**115**(2):e24298

[101] Ciulla M, Di Stefano A, Marinelli L, Cacciatore I, Di Biase G. RNAIII inhibiting peptide (RIP) and derivatives as potential tools for the treatment of *S. aureus* biofilm infections. Current topics in Medicinal Chemistry. 2018;**18**(24):2068-2079

[102] Minich A, Lišková V, Kormanová Ľ, Krahulec J, Šarkanová J, Mikulášová M, et al. Role of RNAIII in resistance to antibiotics and antimicrobial agents in *Staphylococcus epidermidis* biofilms. International Journal of Molecular Sciences. 2022;**23**(19):11094

[103] Khan BA, Yeh AJ, Cheung GY, Otto M. Investigational therapies targeting quorum-sensing for the treatment of *Staphylococcus aureus* infections. Expert Opinion on Investigational Drugs. 2015;**24**(5):689-704

[104] Murray EJ, Crowley RC, Truman A, Clarke SR, Cottam JA, Jadhav GP, et al. Targeting Staphylococcus aureus quorum sensing with nonpeptidic small molecule inhibitors. Journal of Medicinal Chemistry. 2014;**57**(6):2813-2819

[105] Li J, Wang W, Xu SX, Magarvey NA, McCormick JK. *Lactobacillus reuteri*-produced cyclic dipeptides quench agr-mediated expression of toxic shock syndrome toxin-1 in staphylococci. Proceedings of the National Academy of

Sciences of the United States of America. 2011;**108**(8):3360-3365

[106] Nakayama J, Uemura Y, Nishiguchi K, Yoshimura N, Igarashi Y, Sonomoto K. Ambuic acid inhibits the biosynthesis of cyclic peptide quormones in gram-positive bacteria. Antimicrobial Agents and Chemotherapy. 2009;**53**(2):580-586

[107] Qiu J, Jiang Y, Xia L, Xiang H, Feng H, Pu S, et al. Subinhibitory concentrations of licochalcone A decrease alpha-toxin production in both methicillin-sensitive and methicillin-resistant *Staphylococcus aureus* isolates. Letters in Applied Microbiology. 2010;**50**(2):223-229

[108] Khodaverdian V, Pesho M, Truitt B, Bollinger L, Patel P, Nithianantham S, et al. Discovery of antivirulence agents against methicillin-resistant *Staphylococcus aureus*. Antimicrobial Agents and Chemotherapy. 2013;**57**(8):3645-3652

[109] Leonard PG, Bezar IF, Sidote DJ, Stock AM. Identification of a hydrophobic cleft in the LytTR domain of AgrA as a locus for small molecule interactions that inhibit DNA binding. Biochemistry. 2012;**51**(50):10035-10043

[110] Sully EK, Malachowa N, Elmore BO, Alexander SM, Femling JK, Gray BM, et al. Selective chemical inhibition of agr quorum sensing in *Staphylococcus aureus* promotes host defense with minimal impact on resistance. PLoS Pathogens. 2014;**10**(6):e1004174

[111] Bernabè G, Marzaro G, Di Pietra G, Otero A, Bellato M, Pauletto A, et al. A novel phenolic derivative inhibits AHL-dependent quorum sensing signaling in *Pseudomonas aeruginosa*. Frontiers in Pharmacology. 2022;**13**:996871

[112] Savijoki K, San-Martin-Galindo P, Pitkänen K, Edelmann M, Sillanpää A, van der Velde C, et al. Food-grade bacteria combat pathogens by blocking AHL-mediated quorum sensing and biofilm formation. Foods (Basel, Switzerland). 2022;**12**(1):90

[113] Salman MK, Abuqwider J, Mauriello G. Anti-quorum sensing activity of probiotics: The mechanism and role in food and gut health. Microorganisms. 2023;**11**(3):793

[114] Czajkowski R, Jafra S. Quenching of acyl-homoserine lactone-dependent quorum sensing by enzymatic disruption of signal molecules. Acta Biochimica Polonica. 2009;**56**(1):1-16

[115] Li S, Chan KK, Hua MZ, Gölz G, Lu X. Inhibition of AI-2 quorum sensing and biofilm formation in *Campylobacter jejuni* by decanoic and lauric acids. Frontiers in Microbiology. 2021;**12**:811506

[116] Ryu EJ, Sim J, Sim J, Lee J, Choi BK. D-galactose as an autoinducer 2 inhibitor to control the biofilm formation of periodontopathogens. Journal of Microbiology (Seoul, Korea). 2016;**54**(9):632-637

[117] Girennavar B, Cepeda ML, Soni KA, Vikram A, Jesudhasan P, Jayaprakasha GK, et al. Grapefruit juice and its furocoumarins inhibits autoinducer signaling and biofilm formation in bacteria. International Journal of Food Microbiology. 2008;**125**(2):204-208

[118] Bouyahya A, Dakka N, Et-Touys A, Abrini J, Bakri Y. Medicinal plant products targeting quorum sensing for combating bacterial infections. Asian Pacific Journal of Tropical Medicine. 2017;**10**(8):729-743

[119] Srinivasan R, Santhakumari S, Ravi AV. In vitro antibiofilm efficacy of Piper betle against quorum sensing mediated biofilm formation of luminescent *Vibrio harveyi*. Microbial Pathogenesis. 2017;**110**:232-239

[120] Zhang H, Zhou W, Zhang W, Yang A, Liu Y, Jiang Y, et al. Inhibitory effects of citral, cinnamaldehyde, and tea polyphenols on mixed biofilm formation by foodborne *Staphylococcus aureus* and *Salmonella enteritidis*. Journal of Food Protection. 2014;**77**(6):927-933

[121] Kalaiarasan E, Thirumalaswamy K, Harish BN, Gnanasambandam V, Sali VK, John J. Inhibition of quorum sensing-controlled biofilm formation in *Pseudomonas aeruginosa* by quorum-sensing inhibitors. Microbial Pathogenesis. 2017;**111**:99-107

[122] Li J, Shen Y, Zuo J, Gao S, Wang H, Wang Y, et al. Inhibitory effect of monoterpenoid glycosides extracts from peony seed meal on *Streptococcus suis* LuxS/AI-2 quorum sensing system and biofilm. International Journal of Environmental Research and Public Health. 2022;**19**(23):16024

[123] Li J, Fan Q, Jin M, Mao C, Zhang H, Zhang X, et al. Paeoniflorin reduce luxS/AI-2 system-controlled biofilm formation and virulence in *Streptococcus suis*. Virulence. 2021;**12**(1):3062-3073

[124] Yang YB, Wang S, Wang C, Huang QY, Bai JW, Chen JQ, et al. Emodin affects biofilm formation and expression of virulence factors in *Streptococcus suis* ATCC700794. Archives of Microbiology. 2015;**197**(10):1173-1180

[125] Xue B, Shen Y, Zuo J, Song D, Fan Q, Zhang X, et al. Bringing antimicrobial strategies to a new level: The quorum sensing system as a target to control *Streptococcus suis*. Life (Basel, Switzerland). 2022;**12**(12):2006

[126] Ampomah-Wireko M, Luo C, Cao Y, Wang H, Nininahazwe L, Wu C. Chemical probe of AHL modulators on quorum sensing in Gram-negative bacteria and as antiproliferative agents: A review. European Journal of Medicinal Chemistry. 2021;**226**:113864

[127] Waters CM, Bassler BL. Quorum sensing: Cell-to-cell communication in bacteria. Annual Review of Cell and Developmental Biology. 2005;**21**:319-346

[128] Krzyżek P. Challenges and limitations of anti-quorum sensing therapies. Frontiers in Microbiology. 2019;**10**:2473

[129] Wu H, Moser C, Wang HZ, Høiby N, Song ZJ. Strategies for combating bacterial biofilm infections. International Journal of Oral Science. 2015;**7**(1):1-7

[130] Paczkowski JE, Mukherjee S, McCready AR, Cong JP, Aquino CJ, Kim H, et al. Flavonoids suppress *Pseudomonas aeruginosa* virulence through allosteric inhibition of quorum-sensing receptors. The Journal of Biological Chemistry. 2017;**292**(10):4064-4076

[131] Singh V, Evans GB, Lenz DH, Mason JM, Clinch K, Mee S, et al. Femtomolar transition state analogue inhibitors of 5′-methylthioadenosine/S-adenosylhomocysteine nucleosidase from *Escherichia coli*. The Journal of Biological Chemistry. 2005;**280**(18):18265-18273

[132] Huma N, Shankar P, Kushwah J, Bhushan A, Joshi J, Mukherjee T, et al. Diversity and polymorphism in AHL-lactonase gene (aiiA) of *Bacillus*. Journal of Microbiology and Biotechnology. 2011;**21**(10):1001-1011

[133] Park J, Jagasia R, Kaufmann GF, Mathison JC, Ruiz DI, Moss JA, et al.

Infection control by antibody disruption of bacterial quorum sensing signaling. Chemistry & Biology. 2007;**14**(10):1119-1127

[134] Brackman G, Cos P, Maes L, Nelis HJ, Coenye T. Quorum sensing inhibitors increase the susceptibility of bacterial biofilms to antibiotics in vitro and in vivo. Antimicrobial Agents and Chemotherapy. 2011;**55**(6):2655-2661

[135] Zhao X, Yu Z, Ding T. Quorum-sensing regulation of antimicrobial resistance in bacteria. Microorganisms. 2020;**8**(3):425

[136] Fong J, Zhang C, Yang R, Boo ZZ, Tan SK, Nielsen TE, et al. Combination therapy strategy of quorum quenching enzyme and quorum sensing inhibitor in suppressing multiple quorum sensing pathways of *P. aeruginosa*. Scientific Reports. 2018;**8**(1):1155

[137] Givskov M, de Nys R, Manefield M, Gram L, Maximilien R, Eberl L, et al. Eukaryotic interference with homoserine lactone-mediated prokaryotic signalling. Journal of Bacteriology. 1996;**178**(22):6618-6622

[138] Manefield M, de Nys R, Naresh K, Roger R, Givskov M, Peter S, et al. Evidence that halogenated furanones from Delisea pulchra inhibit acylated homoserine lactone (AHL)-mediated gene expression by displacing the AHL signal from its receptor protein. Microbiology (Reading). 1999;**145**(Pt 2):283-291

[139] Balaban N, Novick RP. Autocrine regulation of toxin synthesis by *Staphylococcus aureus*. Proceedings of the National Academy of Sciences. 1995;**92**(5):1619-1623

[140] Ismail AS, Valastyan JS, Bassler BL. A host-produced Autoinducer-2 mimic activates bacterial quorum sensing. Cell Host & Microbe. 2016;**19**(4):470-480

[141] Ma R, Qiu S, Jiang Q, Sun H, Xue T, Cai G, et al. AI-2 quorum sensing negatively regulates rbf expression and biofilm formation in *Staphylococcus aureus*. International Journal of Medical Microbiology. 2017;**307**(4-5):257-267

[142] Siller M, Janapatla RP, Pirzada ZA, Hassler C, Zinkl D, Charpentier E. Functional analysis of the group A streptococcal luxS/AI-2 system in metabolism, adaptation to stress and interaction with host cells. BMC Microbiology. 2008;**8**:188

[143] Xu T, Wang XY, Cui P, Zhang YM, Zhang WH, Zhang Y. The Agr quorum sensing system represses persister formation through regulation of phenol soluble Modulins in *Staphylococcus aureus*. Frontiers in Microbiology. 2017;**8**:2189

[144] Yu D, Zhao L, Xue T, Sun B. Staphylococcus aureus autoinducer-2 quorum sensing decreases biofilm formation in an icaR-dependent manner. BMC Microbiology. 2012;**12**:288

[145] Vuong C, Kocianova S, Yao Y, Carmody AB, Otto M. Increased colonization of indwelling medical devices by quorum-sensing mutants of *Staphylococcus epidermidis* in vivo. The Journal of Infectious Diseases. 2004;**190**(8):1498-1505

[146] He L, Le KY, Khan BA, Nguyen TH, Hunt RL, Bae JS, et al. Resistance to leukocytes ties benefits of quorum sensing dysfunctionality to biofilm infection. Nature Microbiology. 2019;**4**(7):1114-1119

Section 3

Strategies for Eradication of Bacterial Biofilms

Chapter 7

Nanosystems as Quorum Quenchers Targeting Foodborne Pathogens: Understanding the Inhibition Mechanisms and Their Docking Predictions

Dulce María Romero-García,
Jazmín Guadalupe Silva-Galindo,
Carlos Arnulfo Velázquez-Carriles,
Alejandra Monserrat Castorena-Sánchez, Ismael García-Vera,
Gabriela Hinojosa-Ventura and Jorge Manuel Silva-Jara

Abstract

Food poisoning is one of the main problems affecting public health. Bacterial adhesion on surfaces has been documented for decades, and it is known that biofilm-forming bacteria are much more resistant than planktonic cells. Typically, nanosystems are studied regarding their antimicrobial activity (i.e., pathogenic bacteria such as *Campylobacter*, *Salmonella*, *Listeria monocytogenes*, *Escherichia coli* O157:H7, *Staphylococcus aureus*, *Clostridium perfringens*, *Bacillus cereus*, and *Yersinia enterocolitica*), but not for antibiofilm activity and their associated genes. Some studies established protein-ligand prediction concerning quorum sensing suppression, commonly called quorum quenching. This chapter focuses on nanosystems or functionalized nanomaterials that have demonstrated antibiofilm or quorum quenching activity and, thus, establishes perspectives in modeling specific nanosystems to eradicate biofilms produced by foodborne pathogens.

Keywords: nanosystems, quorum quenchers, food safety, biofilm formation inhibition, molecular docking

1. Introduction

Foodborne diseases are a major public health concern, and one factor that has been shown to contribute to their persistence and virulence is biofilm formation by foodborne pathogens. Biofilms are complex structures composed of communities of microorganisms that can attach to surfaces and resist antibiotics and other antimicrobial agents [1, 2].

Many foodborne pathogens, such as *S. aureus*, *Salmonella enterica, L. monocytogenes*, *E. coli* O157:H7, and *Vibrio*, can form biofilms. Biofilm formation by foodborne pathogens can occur in various settings, including food processing facilities and in the human body during infection. Their capacity to form biofilms can contribute to prolonged contamination of food processing surfaces, leading to outbreaks and significant economic losses. In addition, biofilms can provide a protective environment for the bacteria in the human body, making them difficult to eradicate and potentially leading to recurrent infections [3].

Efforts to control biofilm formation by foodborne pathogens are important for reducing the risk of foodborne disease. Strategies to prevent and mitigate biofilm formation include proper cleaning and disinfection of food processing surfaces, targeted use of antimicrobial agents, and development of new drugs that can disrupt biofilm formation.

1.1 Biofilm formation

The biofilm formation process differs among bacteria, but the process generally involves several stages [4]. Biofilms are formed by extracellular polymeric substances (EPS) that may be composed of proteins, DNA, and polysaccharides [5, 6]. Some bacteria can form this type of biofilm in situations of change or stress. A bacterial community is formed with the ability to adhere to inert materials or living tissues. In this microenvironment, the bacteria among themselves carry out a type of communication, or chemical signaling, that consists of the production of inducing molecules [communication by autoinducers (AI)]. This type of signaling mechanism is known as "quorum sensing.". Quorum sensing (QS) communication occurs not only between bacteria, but they can also associate a microorganism different from the initial species through the biofilm; the other microorganisms will have the same genes that they express for biofilm formation and thus be able to also resist the stress that initiated the mutation, as shown in **Figure 1** [7, 8].

Different structures and proteins are related to biofilm formation and stabilization, whether Gram-positive or -negative bacterium is involved. These include lipopolysaccharides, exopolysaccharides, QS, teichoic acids, and others [9, 10].

The specific genes associated with QS and biofilm formation can vary between foodborne pathogen species and strains; some examples are summarized in **Table 1**.

1.2 Quorum sensing and quorum quenchers

Quorum sensing, a process of cell-to-cell communication used by many bacteria to regulate their behavior, has been identified as a potential target for combating foodborne pathogens. Quorum quenchers (QQs) are compounds that can interfere with QS, representing a promising strategy to control bacterial infections. In recent years, the use of nanotechnology-based approaches, such as green synthesis of nanoparticles, has been explored for synthesizing alternative QQs agents that can be used against foodborne pathogens. Additionally, with the advent of computational chemistry and molecular docking techniques, it has become possible to develop computational models that can predict the interactions between QS and QQs.

The use of quorum quenchers has been an attractive approach to inhibit the virulence and biofilm formation of bacteria, without killing them, thus reducing the

DOI: http://dx.doi.org/10.5772/intechopen.112266

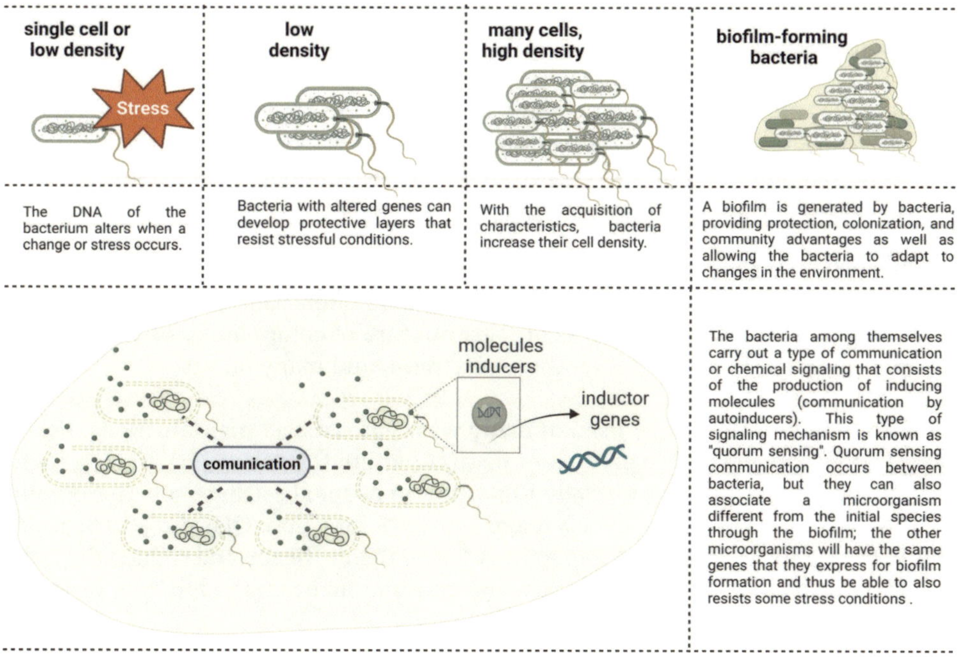

Figure 1.
Biofilm formation and quorum sensing bacteria.

Gene	Microorganism associated	Function
*lux*I and *lux*R	*Vibrio fischeri*, *E. coli*, *S. enterica*	Production and detection of quorum sensing signaling molecule acyl-homoserine lactone (AHL).
*las*I and *las*R	*P. aeruginosa*	Control production and detection of N-(3-oxododecanoyl)-L-homoserine lactone (3-oxo-C12-HSL).
*rhl*I and *rhl*R	*P. aeruginosa*	Control production and detection of N-butyryl-homoserine lactone (C4-HSL).
*agr*A and *agr*C	*S. aureus*	Control production and detection of autoinducing peptide (AIP).
*lux*S	Various	Production of autoinducer-2 (AI-2).
csg and *bcs*	*E. coli*, *S. enterica*, *L. monocytogenes*	Production of curli and cellulose, respectively, which are essential components of the biofilm matrix.
*rpo*S	*S. enterica*, *E. coli*	Encodes for the RpoS protein, involved in the regulation of stress response and the formation of persistent cells in biofilms.

Table 1.
Genes associated with quorum sensing in Gram-positive and negative bacteria [11, 12].

risk of resistance development. There are different types of QQs, such as enzymatic and non-enzymatic, based on different mechanisms of action:

1. *Enzymatic QQs*: These are enzymes that specifically degrade the signaling molecules that mediate QS. Examples include lactonases, acylases, and oxidoreductases. These enzymes can hydrolyze the acyl-homoserine lactones

(AHLs) or peptides, the most common QS molecules, into inactive forms that cannot activate the QS machinery.

2. *Non-enzymatic QQs*: These are small molecules that can bind to QS signaling molecules and prevent them from binding to their receptors or even degrade them. Examples of non-enzymatic QQs include furanones, halogenated furanones, or quinolones.

3. *Natural QQs*: Some phytocompounds or bacteria have evolved quorum quenching mechanisms that help them regulate interactions with other bacteria. These natural QQs comprise a complex mixture of compounds, such as essential oils, gingerol, kahweol, resveratrol, curcumin, and many others.

Quorum quenchers can be used in many ways against bacterial infections. For example, they can be used to prevent or disrupt biofilm formation on surfaces, such as medical equipment. This is especially important in hospital settings, where microbial biofilms on medical devices can be a major source of infection. QQs can also be used in combination with antibacterial agents to enhance their efficacy and reduce the risk of resistance development. For instance, using quorum quenchers to mitigate resistance development can be useful in chronic infections.

1.3 Nanosystems

Nanoparticles (NPs) are particles that have at least one dimension between 1 and 100 nanometers. They can be naturally occurring or artificially created and have unique physical and chemical properties that differ from those of their bulk counterparts. They are studied in many scientific fields such as physics, chemistry, geology, and biology. NPs have many applications in industry and medicine, such as in drug delivery systems, in targeted cancer therapies, and in developing new materials and electronics.

Antimicrobial nanoparticles are types of nanomaterials that have shown potential for use as antimicrobial agents due to their unique physicochemical properties. They can be synthesized from a variety of materials including metals, metal oxides, and polymers. These NPs can be used to inhibit the growth of different microorganisms such as bacteria, viruses, and fungi, either by disrupting their cell membranes or by interfering with their metabolic processes. Additionally, antimicrobial nanoparticles have been studied for use in applications such as food packaging, wound dressings, and water treatment.

Nanosystems in biological science typically refer to systems or structures at the nanoscale that are relevant to the study of biology. Nanosystems have emerged as a promising approach for the prevention of biofilm formation and antimicrobial resistance.

Nanosystems have several mechanisms that can prevent biofilm formation and overcome antimicrobial resistance. These mechanisms include the physical disruption of biofilm, targeting biofilm matrix, preventing bacterial adhesion, and releasing antimicrobial agents in a controlled and sustained manner. In addition, nanosystems can improve the pharmacokinetics and pharmacodynamics of antimicrobial agents, enhancing their efficacy and reducing their toxicity.

Nanosystems have also demonstrated great potential in the development of novel antimicrobial agents that can overcome antimicrobial resistance mechanisms. For

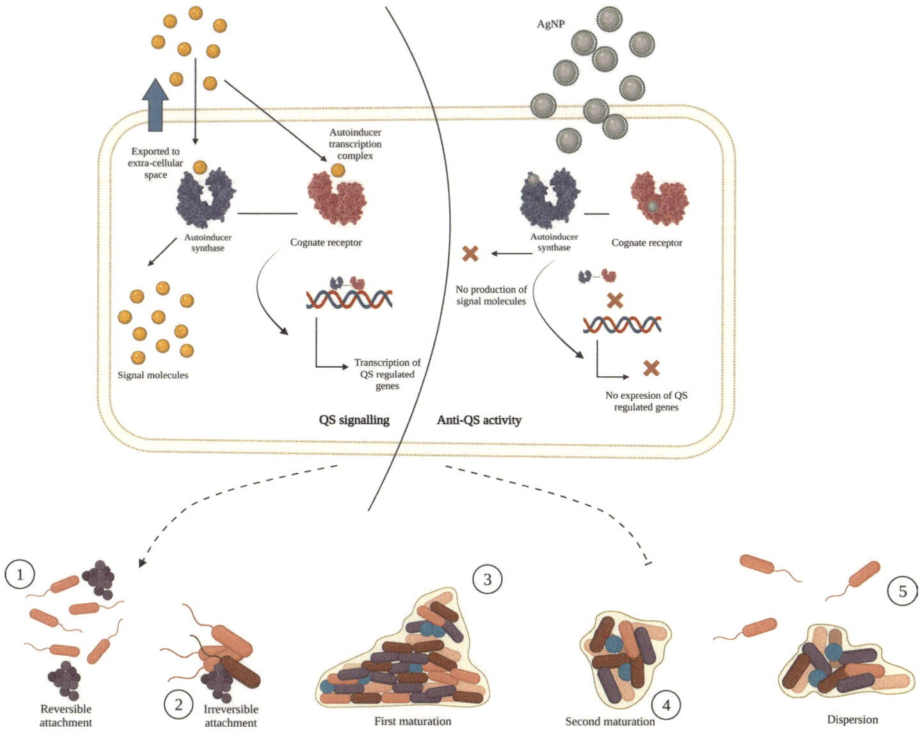

Figure 2.
AgNPs QS signalling and anti-QS mechanism in biofilm formation in bacteria.

example, the use of silver nanoparticles (Ag-NPs) has been reported to be effective against several drug-resistant bacteria, including methicillin-resistant *S. aureus* (MRSA) and *Pseudomonas aeruginosa*. In **Figure 2**, we can observe the mechanism by which QS and anti-QS activity occur in bacteria, leading to the formation (dashed arrow, left) or inhibition of biofilm (dashed line, right) throughout the five stages of the bacterial biofilm formation process.

However, the development of nanosystems as a viable alternative to traditional antibiotics requires careful consideration of their potential toxicity, biocompatibility, and long-term safety.

1.4 Molecular docking

Molecular docking is a computational method used to study the interactions between a protein and a small molecule, called a ligand. The goal is to predict how the ligand will bind to the protein and the strength of that interaction.

The process of molecular docking involves several steps, including preparing the protein and ligand structures, generating a docking grid, and running the docking simulation. The simulation predicts the optimal docking position and conformation, with the goal of optimizing a scoring function. After running the simulation, the result is analyzed to better understand the protein-ligand interactions and assess the potential binding strength. The result is then validated using experimental data or benchmarking.

Recently, molecular docking has been explored for its potential applications in biofilm formation control by targeting QS systems of bacteria. By understanding how small molecules interact with key components of QS pathways, we can develop strategies to disrupt these pathways and thus prevent biofilm formation without harming other beneficial bacterial species or human health concerns associated with antibiotic use.

In recent years, several studies have employed molecular docking as a tool for the design and development of novel antibiofilm agents. For example, a study by Khadke, see [13], used molecular docking to evaluate the binding affinity of cinnamaldehyde analogs with the biofilm response regulator yeast-form wall protein 1 (YWP1) and upregulated by cAMP in filamentous growth (UCF1) in *Candida albicans*. The study demonstrated that the cinnamaldehyde analogs could effectively inhibit biofilm formation in the yeast. Moreover, molecular docking can assist in identifying the interaction mechanism of an antibiofilm agent with its target molecule. For instance, a study by Ren [14] showed that isookanin could bind to biofilm-related proteins and interrupt biofilm formation in *Staphylococcus epidermis*.

For this, molecular docking is a promising technique for the design and development of novel antibiofilm agents. The use of molecular docking could assist in identifying the interaction mechanism of an agent with its target molecule and predict the therapeutic efficacy of the agent. However, further studies are necessary to validate the outcomes of molecular docking through *in vitro* and *in vivo* experiments.

2. *P. aeruginosa*

P. aeruginosa is an opportunistic Gram-negative bacterium associated with nosocomial diseases, mainly lung and airways infections. One of the main mechanisms of action relates to using degradation enzymes and biofilm formation. The biofilm produced by *P. aeruginosa* gives it a sessile behavior making it difficult to attack by antimicrobial agents due to its exopolysaccharide nature and persistence of virulence factors [15, 16].

The genus *Pseudomonas* spp. is involved in the colonization and management of processes in the dairy industry in which different temperatures are used, allowing to take advantage of these conditions and some food structures, as surface substrates for biofilm formation [1].

2.1 Quorum sensing and related genes

Bacterial growth of *P. aeruginosa* has shown structural and metabolic changes ranging from the planktonic phase to the sessile or cellular attachment phase, identifying up to five stages in which biofilm formation takes place. These stages are: (1) reversible attachment, (2) irreversible attachment, (3) first maturation, (4) second maturation, and (5) dispersion, where each of them exhibits the expression of different protein patterns [17, 18].

In Gram-negative bacteria, AHLs have been characterized as the main molecules triggering QS signaling. In the QS of *P. aeruginosa*, several genes associated with each of the stages of biofilm formation have been identified. Signaling and regulatory genes include: N-3-oxododecanoyl *L*-homoserine lactone (3-oxo-C12-HSL), N-butyryl-homoserine lactone (C4-HSL), 2-heptyl-3-hydroxy-4-quinolone, and 2-(2-hydroxyphenyl)thiazole-4-carbaldehyde, controlled by the regulatory systems such as *las*I/R, *rhl*I/R, *pqs*ABCDE/*pqs*R, and *Amb*BCDE/*Iqs*R, respectively. The synthesis of these AI improves protein expression leading to an increase of factors involved in QS. Also, maturation of the biofilm is regulated by this feature, following irreversible attachment, production of virulence factors involved in *P. aeruginosa* pathogenicity, iron scavenging activity, motility, and dissemination [19, 20].

Biosynthesis inhibition of AI molecules can be biodirected through modified nanosystems as an alternative to diminish the virulence of *P. aeruginosa.*

2.2 Nanosystems tested *in vitro* and *in silico* for biofilm inhibition

In effort to inhibit QS mechanisms of opportunistic bacteria like *P. aeruginosa*, green alternatives based on nanosystems have been developed. **Table 2** exhibits nanosystems synthesized using plant extracts, microorganisms, or bioactive coatings, aiming to inhibit biofilm formation and QS activity of *P. aeruginosa*. In 2017 [24], Elshaer and Shaaban proposed selenium nanoparticles (Se-NPs) coated with honey polyphenols used as nanovectors in drug delivery systems, suggesting molecular docking studies as demonstration of antivirulence potential against *P. aeruginosa*.

Additionally, antibiofilm activity of 68% was exposed for Au-NPs, with a quorum quencher activity up to 88% at 4.6 $\mu g \cdot mL^{-1}$, using synthesis of NPs by biological reduction with different strains of *Streptomyces* isolated from soil. The reduction of selenium and gold metal ions to nanometals (Se-NPs and Au-NPs) was carried out

Nanosystem	Antibiofilm activity assay %	Quorum quencher activity (% or concentration)	Ligands	Receptor protein (genes)	Reference
Se-NPs Au-NPs	64–88% with 2.3 $\mu g \cdot mL^{-1}$ 26–68% with 4.6 $\mu g \cdot mL^{-1}$	65–90% with 2.3 $\mu g \cdot mL^{-1}$ 40–88% with 4.6 $\mu g \cdot mL^{-1}$	Pyocyanin Elastase Protease	*las*R *las*I *rhl*I *rhl*R *pqs*A *pqs*R *tox*A *las*A *las*B	[21]
ZnO nanospikes	20–85% with 25–200 $\mu g \cdot mL^{-1}$	35–75% with 25–200 $\mu g \cdot mL^{-1}$	Elastase Protease Pyocyanin	*las*B	[22]
Piper betle-Ag-NPs	78.20% with 8 $\mu g \cdot mL^{-1}$	82.43% with $\mu g \cdot mL^{-1}$	Eugenol Pyocyanin Elastase	*las*I *las*R *Mvf*R	[23]
Se-nanovectors coated with honey	>90% with 4.5 $\mu g \cdot mL^{-1}$	100% 4.5 $\mu g \cdot mL^{-1}$	Acacetin Apigenin Pinocembrin Pinobanksin Quercetin Caffeic acid Kaempferol	*las*R	[24]

Table 2.
Biofilm formation inhibitors nanosystems tested in vitro *and* in silico *against* P. aeruginosa.

under green chemical conditions monitoring the antibiofilm activity by crystal violet method [21]. The anti-QS and antibiofilm activity of zinc oxide (ZnO) nanospikes coated with cetyltrimethylammonium bromide (CTAB) was shown by Prateeksha [22], with different incubation times. The antibiofilm activity of *P. aeruginosa* was determined using the polyvinyl chloride biofilm formation assay and crystal violet cell attachment assay, in addition to elastase and protease transcriptional activity analysis for QS [22, 25]. About aqueous plant extracts, a bioreduction of Ag-NPs mediated by *Piper betle* (*Pb*) leaves was evaluated using molecular docking of the interaction of NPs conjugated with Eugenol (the main phenolic compound of *Pb* leaves) and QS-associated proteins of *P. aeruginosa* [23]. Results revealed an antibiofilm activity of 78% and a quorum quencher activity of 82% at 8 $\mu g \cdot mL^{-1}$ concentration, implying considerable interactions between Eugenol-Ag NPs and QS-regulatory proteins.

The characteristics steered by each of these biologically mediated nanosystems provide a perspective for green molecular strategies targeting microorganisms such as those developed in this chapter.

3. *Salmonella*

Salmonella species are a group of Gram-negative bacteria, which causes animal and human infections. *Salmonella* genus contains two species, *S. enterica* and *Salmonella bongori*, the first being subdivided into six subspecies [26]. Based on the clinical syndromes that *Salmonella* spp. cause, it could be classified as typhoid *Salmonella* and non-typhoid *Salmonella* (NTS). Only Typhi and Paratyphi serotypes are causative agents of typhoidal fever, an acute illness with symptoms that include high fever, malaise, and abdominal pain; typhoid fever has been associated with 600,000 deaths per year [27]. NTS serotypes cause gastroenteritis that is typically uncomplicated, however, it can be severe for immunosuppressed patients, elderly, and infants. NTS illnesses are the fourth morbidity and the third leading cause of deaths among diarrheal diseases worldwide. Every year, *Salmonella* spp. causes 93 million cases of gastroenteritis and more than 150,000 deaths, among these, 85% of cases were food linked [28–30]. The most prevalent serovars of *Salmonella* are Enteritidis, Newport, Typhimurium, and Javiana [31, 32].

3.1 Quorum sensing and related genes

During biofilm formation, microorganisms can communicate with each other through QS to regulate metabolic activity. QS is mediated by three mechanisms of autoinducers: AI-1, AI-2, and AI-3. The main regulators of pathogenic *Salmonella* are AI-2 and AI-3 [33].

Salmonella produces AI-2 through *luxS* gene, and SdiA protein detects AHLs produced by other bacterial species, with preference of oxoC18 modification; however, it can detect AHLs with other structures such as oxoC12 produced by *P. aeruginosa*, C4 produced by *Aeromonas hydrophila*, or C6 and oxoC6 produced by *Y. enterocolitica* [34, 35]. AI-2 signaling requires low pH and high osmolality, as low osmolality induces signal degradation. The AI-3 synthetic pathway and chemical formula remain unknown. Despite this, epinephrine, norepinephrine, and catecholamines are associated with AI-3 regulation [35].

Many genes are associated with *Salmonella* biofilm formation, like *lux*S/AI-2/*lux*R homolog SdiA system related to QS, the QseBC two-component system also associated with QS, and a universal regulator of virulence [34]. Other genes associated with biofilm formation are *Mig-14* and *Virk* genes, which reduce outer membrane permeability and induce polymyxin B resistance, *Mig-14* gene is stimulated by antimicrobial peptides and acidic pH conditions [36]. *S. enteritidis* and *S. typhimurium* have the *rck* gene on virulence plasmids, inducing cellular adhesion and increasing bacterial resistance to serum [37].

3.2 Nanosystems tested *in vitro* and *in silico* for biofilm inhibition

Many strategies have been developed to achieve biofilm and QS inhibition, including *in vitro* and *in silico* tests. A virtual screening was performed by Almeida, see [38], for nonsteroidal anti-inflammatory drugs, searching potential inhibitors of QS in *Salmonella* using molecular docking. This study considered three different macromolecular arrangements of SdiA protein observing binding affinities testing more than 193 compounds. Z-phytol and lonazolac molecules were recognized as candidates for *in vitro* inhibition. Also, *in silico* anti-QS activities of Benzeneethanamine, 4-methoxy- and 2-Cyclopentadecanone, 2-hydroxy by a SdiA protein interaction were predicted [39]. Phytochemical berberine was studied [40] as an antibiofilm inhibitor using crystal violet microtiter plate assay; berberine showed 31.20% antibiofilm activity at 0.019 mg·mL^{-1} in front *S. enterica* sv Typhimurium and QS inhibitory potential was screened using Chromobacterium *violaceum* biosensor bacteria. Inhibitors of biofilm and Quorum quenchers are presented in **Table 3**, using plant-based molecules tested *in vitro* and *in silico*.

Nanosystem	Antibiofilm activity assay %	Quorum quencher activity (% or concentration)	Ligands	Receptor protein (genes)	Reference
Plant compounds and nonsteroidal anti-inflammatory drugs	Not specified, indicated as high activity in 83.20% of the compounds	†	193 compounds anti-inflammatory and plant compounds	SdiA (4Y13-S, 4Y15-S, 4Y17S)	[38]
Phytocompounds of *Psidium guajava*	†	†	Benzeneethanamine, 4-methoxy- and cyclopentadecanone, 2-hydroxy-	SdiA	[39]
Berberine	31.20% with 0.625 mg mL^{-1}	†	Berberine	*las*R	[40]
Lactic acid Malic acid	†	80.2% in spinach 76.6% in cantaloupe 46.7% in spinach 37.5% in cantaloupe 80.4% in cantaloupe	Lactic and malic organic acids	*lux*S	[41]
Ag-NPs of *Myristica fragrans* seed extract	87% nutmeg aqueous seed extract 99.1% with 50 μg·mL^{-1} Ag-NPs	†	6,6a-dihydro-1-(1,3-dioxolan-2-yl)-, (3aR, 1-t, octadecane, 6-methyl-, heptadecane, 2,6,10,14-tetramethyl-BIS (2-ethylhexyl) phthalate	RcsB RcsC	[42]

†Not declared.

Table 3.
Biofilm formation inhibitors nanosystems tested in vitro *and* in silico *against* Salmonella *spp.*

4. *E. coli*

E. coli is a Gram-negative, rod-shaped bacterium from the *Escherichia* genus. Different serotypes that produce toxins are associated with foodborne diseases, such as *E. coli* O157:H7, commonly known as enterohemorrhagic *E. coli* (EHEC). This pathogen is responsible for bloody diarrhea outbreaks and hemolytic uremic syndrome worldwide [43].

Virulence of EHEC is well studied, and it involves different mechanisms through which pathogen survives the acid environment in the stomach of the host, and colonizes the intestine, where lesions are provoked and, consequently, bloody diarrhea occurs [44].

4.1 Quorum sensing and related genes

Quorum sensing of *E. coli* is mediated by the *lux*I/*lux*R system, where AI are synthesized and recognized; as a result, different phenotypes, such as bioluminescence, antibiotic production, biofilm formation, and virulence factors' secretion, are expressed [45]. The most common AI in *E. coli* are AHLs, and some of these lactones have different lengths of acyl tail (4–20°C), oxidation at the third carbon in the acyl tail (carbonyl, alcohol, or methylene), units of unsaturation, or aryl located opposite to aliphatic tails [45, 46].

Another receptor found in *E. coli* is the SdiA receptor, for which the specific AHL is not present in the genome of this pathogen. Instead, this receptor can interact with AHLs produced by other bacteria that exist in the host, favoring the expression of the *gad* operon, which includes proteins associated with acid environment resistance, a crucial step in the colonization and infection of *E. coli*. In the intestine, SdiA-AHL complex is dissociated, leading to the activation of the LEE operon (locus of enterocyte effacement), which is related to lesions provoked on the walls of the intestine and bloody diarrhea [45].

The disruption of the QS has been one of the most studied methods to control growth and virulence of foodborne pathogens, and some authors suggest different pathways to achieve this disruption [46]:

- Inhibition of AI synthesis.
- Enzymatic degradation of AI.
- Use of materials that adsorb or quench AI.
- Use of compounds that mimic AI affinity for the proteins related to recognition of AI.

4.2 Nanosystems for QS inhibition

Inhibition of QS has gained more attention nowadays, since AHLs can go in and out of a cell, different molecules have been proposed to interact with the specific receptors, mimicking the AHL structure or blocking the receptors. In **Table 4**, some examples of QS inhibitors and their proposed mechanisms are compiled.

Nanosystem	Synthesis method	Concentration	Mechanism	Reference
Chitosan (CS)-NPs with quercetin	Ionotropic gelation	†	Quercetin binds to different domains of LuxR receptor, affecting binding affinity to LuxI-DNA. Other suggestion is that quercetin blocks the secretion of AHL to the cytosol.	[47]
CS Nanoemulsion/ Nanocapsule doped with cinnamaldehyde	Spontaneous emulsification	0.5 mM	Green fluorescence protein (GFP) inhibition.	[46]
D-limonene nanoemulsion	Spontaneous emulsification	†	AI synthesis and motility reduction, suppression of extracellular polymeric substances (biofilm formation inhibition).	[48]
CS-NPs dually crosslinked with genipin and sodium tripolyphosphate (TPP)	Ionotropic gelation	0.001–0.1 $mg{\cdot}mL^{-1}$	AHL-mediated fluorescence response decrease.	[45]
CS nanosystem functionalized with mannose	Reductive amination reaction	3 $mg{\cdot}mL^{-1}$	Interference with lectins, avoiding adhesion, motility, and biofilm formation.	[49]

†Not declared.

Table 4.
Nanosystems and their mechanisms of E. coli *QS inhibition.*

4.3 Molecular docking applied in *E. coli*

Several researches have modeled the possible interaction of nanosystems of bioactive compounds on specific genes or proteins expressed by *E. coli*. For example, synthetic thiazolo[3,2-α]pyrimidine molecules were obtained using ZnO nanoparticles as catalyst, and the *in silico* test was used to determine with which residues from the DNA gyrase B they bind. The results are shown in **Table 5**.

According to Schembri [51], several genes are affected in expression during biofilm formation, growth, and stationary phases. Among them, it has been shown that *flu* and *rpo*S are two of the most important genes for *E. coli* biofilm formation. *Flu* expresses the formation of the antigen 43 (Ag43), which deals with autoaggregation of cells, a primary step for biofilm to begin formation, while *rpo*S activates other genes in charge of dealing with stress conditions such as carbon starving, oxidative degradation of DNA, osmotic stress, etc.

Zinc oxide nanoparticles obtained with green synthesis using *Dysphania ambrosioides* extract were evaluated for molecular docking against *E. coli*. It was demonstrated that ZnO nanoparticles interact with the AcrAB-TolC proteins, which is a pump that crosses the inner and outer membrane of Gram-negative bacteria. This protein has been associated with resistance to antibiotics development, for which inhibiting its expression may lead to growth control of the pathogen [52].

In other cases, the extracts used in green synthesis of nanoparticles are studied for molecular docking. Such is the case of *Aegle marmelos* extract used to obtain copper oxide nanoparticles, and which was estimated to show that the major components

Molecule	Docking score (kcal·mol^{-1})	H-bonds	Amino acid residual interaction	
			Hydrophobic/ Pi-cation/ Pi-anion/ Pi-alkyl interactions	Van der Waals interaction
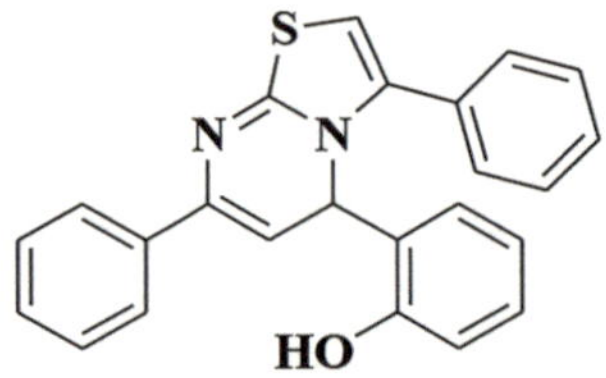	−7.4	Asn-46, Thr-165	Val-43, Ala-47, Gly-77, Ile-78, Ile-94, Val-167	Asp-49, Glu-50, Asp-73, Arg-76, Pro-79
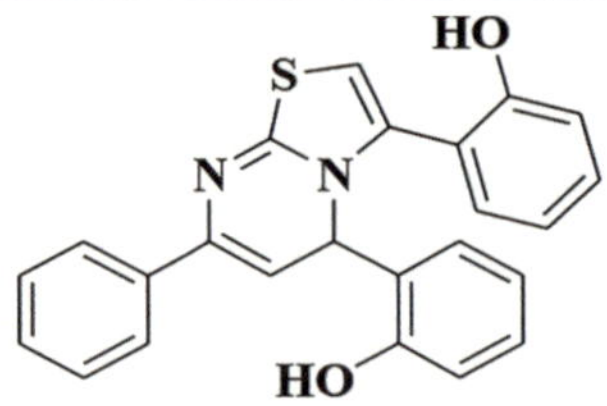	−7.6	Gly-77, Thr-165	Val-43, Ala-47, Asp-73, Ile-78	Pro-79
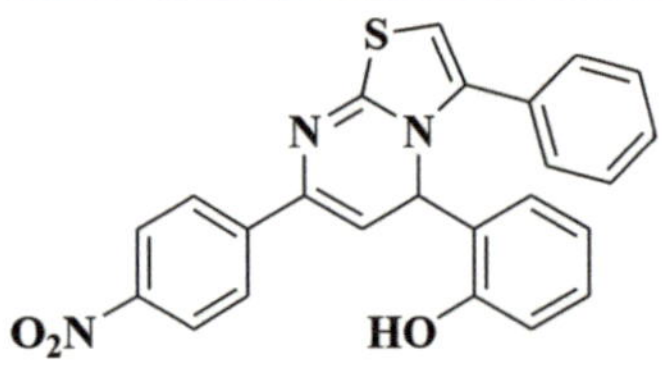	−7.4	Ser-121	Ile-78, Pro-79, Ile-94	Val-97, Leu-98, Gly-119, Val-120
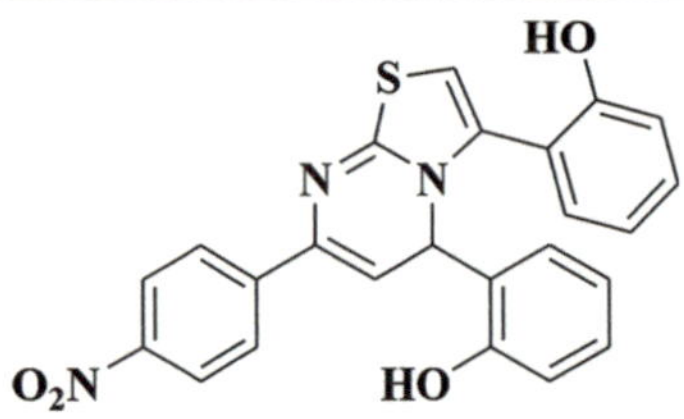	−7.3	Glu-50	Arg-76, Ile-78, Pro-79	Ala-47, Asp-73, Gly-77, Thr-165
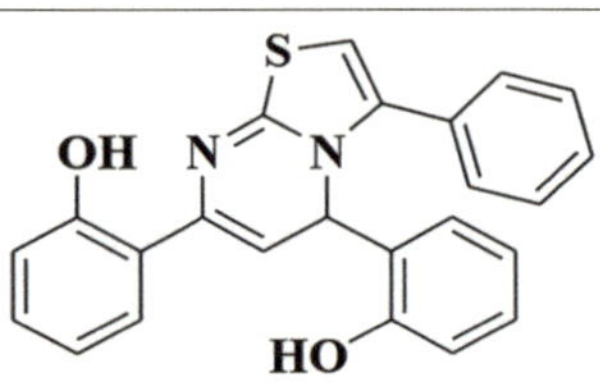	−7.6	Asn-46	Val-43, Ala-47, Asp-73, Gly-77, Ile-78, Ile-94, Thr-165, Val-167	Val-71, Val-120

Table 5.
Bound residues of E. coli *DNA gyrase B of synthetic thiazolo[3,2-α] pyrimidines [50].*

were beta-sitosterol, gamma-sitosterol, and marmesin. These compounds were studied *in silico* against the BamA protein of *E. coli*, a pump that allows substrates to insert into the outer membrane of Gram-negative bacteria. Results showed binding energies of −8 kcal·mol^{-1}, except for gamma-sitosterol (12 kcal·mol^{-1}). Interaction of these compounds present in the extract suggests good inhibition of the protein. Beta-sitosterol forms both polar and non-polar bonds like hydrogen, pi-sigma, pi-alkyl bonds, and Van der Waals interaction with the residues GLU435, PHE494, TYR653, and ASN666 of the BamA; on the other hand, gamma-sitosterol binds only to the ASN534 and TYR468. Both present a stable complex. Finally, marmesin binds with

DOI: http://dx.doi.org/10.5772/intechopen.112266

ARG734, THR588, PHE586, and ARG583 with hydrogen bonds, while non-polar interactions can happen [53].

5. *S. aureus*

Staphylococci are spherical, non-sporulating, Gram-positive bacteria that are found in irregular grape-like clusters. The genus *Staphylococcus* is comprised of at least 45 species, four of which, *S. aureus*, *Staphylococcus epidermidis*, *Staphylococcus lugdunensis*, and *Staphylococcus saprophyticus*, are considered the most important in clinical terms [54]. *S. aureus* is considered an opportunistic pathogenic bacterium, causing a considerable number of diseases ranging from superficial skin and soft tissue infections (SSTI) to invasive infections, sepsis, and death. So, in the United States, the mortality rate due to *S. aureus* sepsis has only been about 20,000 deaths per year in recent years [55]. Therefore, this bacterium has created a resistance to the best available antistaphylococcal agents, such as penicillin and methicillin. However, researchers have developed strategies to inhibit the quorum sensing control of *S. aureus*, since the toxins, virulence factors, and biofilm formation of this pathogen are controlled by the Agr (accessory gene regulator) quorum sensing system [55, 56].

5.1 Quorum sensing and related genes

The expression of the virulence factors of *S. aureus* is controlled by the Agr-QS system, which is responsible for causing genetic adaptations for intracellular communication. Likewise, the Agr of *S. aureus* is characterized mainly by regulating the expression of different toxins, virulence factors and controlling the interaction of bacteria-host at the infection site [57, 58]. Like many other bacterial physiological functions, the formation of *S. aureus* biofilms is mainly encoded by 12 different genes such as the fibrinogen-binding protein (Fib) gene, the fibronectin-binding protein (FnbA and FnbB) genes, intercellular adhesion genes (*ica*A, B, C, and D), clumping factor (*clf*A and B), elastin-binding protein (elastin-binding protein of *S. aureus* (EbpS)), laminin-binding protein (Eno), and collagen-binding adhesin protein (Cna) [59].

The genes encode different surface proteins that allow *S. aureus* to adhere to, penetrate into, and colonize the host. Ultimately, it leads to biofilm formation and virulence. In the *S. aureus* biofilm, the *fib* gene is responsible for facilitating and encoding the recognition of surface fibrinogen-binding proteins, while Cna promotes adhesion to the surface. For their part, the intracellular adhesion genes *ica*A, B, C, and D encode the process of cell-to-cell adhesion and start the formation of biofilms. While the clumping factor genes *clf*A and *clf*B encode cell wall-anchored proteins that bind to host surface fibrinogen [59, 60], facilitating *S. aureus* colonization, biofilm formation, and eliciting virulence via immune evasion through the binding of soluble fibrinogen.

5.2 Nanosystems tested *in vitro* and *in silico* for biofilm inhibition

Researchers have been seeking different strategies to counter *S. aureus* virulence factors, since virtually many of the toxins and other *S. aureus* virulence factors are controlled by the Agr quorum sensing system. For this reason, scientists have dedicated themselves to investigating strategies that manage to inhibit the QS control of *S. aureus* [61]. **Table 6** shows some studies of nanosystems synthesized from different precursors, which aim to inhibit the formation of *S. aureus* biofilms.

Nanosystem	Synthesis process	Chemical precursor	Antibiofilm activity assay %	Receptor protein (genes)	Reference
Graphitic carbon nitride nanosheets decorated with Cu-NPs	Cu-NPs: Cu precipitation and sonication. Carbon nitride nanosheets: melamine heating at 550°C for 4 h followed by precipitation.	$Cu(NO_3)_2$, SDS, L-ascorbic acid, and $C_3H_6N_6$	65% at 312.5 $\mu g \cdot mL^{-1}$	*ica*A	[62]
CS/Ag nanocomposite	Precipitation and heating at 90°C for 6 h	Ag^+ as precursor and CS as stabilizing reducing agent	96% at 250 $\mu g \cdot mL^{-1}$	†	[63, 64]
CS-ZnO-gentamicin nanocomposite	Precipitation and drying of NP at 60°C	$Zn(NO_3)_2$	77% at 0.125 $\mu g \cdot mL^{-1}$	†	[65]

†Not declared.

Table 6.
Biofilm formation inhibitors nanosystems tested in vitro *and* in silico *against* S. aureus.

Biofilm inhibition in the *S. aureus* isolate was shown to be independent of the *ica*A gene, leading to biofilm inhibition in *S. aureus* after treatment with Cu/g-C_3N_4 nanocomposites. On the other hand, Ramachandran [63] demonstrated that CS-loaded Ag-NPs favorably inhibited the formation of *S. aureus* biofilm at a concentration of 250 $\mu g \cdot ml^{-1}$. Therefore, they confirmed that there were damages in bacterial growth, arrest of survival, deformation of the membrane, and alterations of the exopolysaccharide when increasing the concentration of silver nanocomposites loaded with chitosan (CS). Finally, Hemmati [65] demonstrated that incorporating gentamicin-loaded ZnO-NPs into a chitosan solution developed a slow drug release rate compared to gentamicin-conjugated CS-ZnO NPs. Likewise, with the three components (gentamicin, chitosan, and ZnO), the scientists showed that the greatest antimicrobial and antibiofilm activity against *S. aureus* and *P. aeruginosa* occurred in the gentamicin-loaded CS-ZnO nanocomposite, due to the synergistic action that presented the gentamicin with the nanocomposite.

6. *L. monocytogenes*

Listeriosis is a serious infection that is usually caused by eating food contaminated with *Listeria*. In the United States, approximately 1600 people contract listeriosis each year, and approximately 260 people die from the disease [66]. Additionally, mortality from this infection can be as high as 30% in some parts of the United States. In European countries, the European Food Safety Authority (EFSA) reported a total of 1760 cases of listeriosis in humans in 2013 [67]. Contaminated ready-to-eat foods, such as soft cheeses made primarily from unpasteurized milk, smoked fish, ice cream, melon, apple, and vegetables, have been implicated in *L. monocytogenes* [68].

L. monocytogenes is a Gram-positive bacterium with a diameter of 0.5 to 4 μm and a length of 0.5 to 2 μm. It is a facultative anaerobic, catalase-positive, oxidase-negative, non-spore-forming microorganism. It is generally motile due to the presence of flagella in a temperature range of 22–28°C, but immobile above 30°C. The growth temperature of *L. monocytogenes* is −0.4 to 45°C, with an optimum temperature of 37°C. The bacterium can survive in water activity <0.90 and pH 4.6–9.5 and tolerate salt levels (NaCl)) up to 20%. Furthermore, *L. monocytogenes* is resistant to disinfectants and can adhere to different surfaces [69, 70].

6.1 Quorum sensing and related genes

Biofilm formation of *L. monocytogenes* can be influenced by several external factors, such as growth and pressure conditions, temperature, growth method, physicochemical properties of the substrate, and the presence of other microorganisms [71], as well as internal factors such as *prf*A, *act*A, proteins encoded by the σ^B gene, and the ABC (ATP-binding cassette) permease transporter gene [72]. The *L. monocytogenes* genes involved in flagellar motility (*fliQ*, *fla*A, *fli*1, *mot*A) are required for biofilm formation, such as the PhoR gene (phosphate sensory operon) and the genes involved in D-alanine uptake in lipoteichoic cells [1].

L. monocytogenes is sensitive to a wide range of antibiotics active, except cephalosporins and fosfomycin, to which it has inherent resistance. The most common treatment for listeriosis is ampicillin or a combination of ampicillin with gentamicin; however, the *fos*X, *lin*, *abc-f*, and *tet*(*M*) genes are the four most common antimicrobial resistance genes found in *L. monocytogenes* in cases of foodborne transmission [73].

6.2 Nanosystems tested *in vitro* and *in silico* for biofilm inhibition

The reduction and elimination of biofilm of *L. monocytogenes* have been studied through the synthesis and application of nanoparticles and nanosystems. In addition, this nanotechnology can be green technology (**Table 7**).

Synthesized ZnO nanostructures from *Nigella sativa* seed affect biofilm without influencing the bacterial growth, resulting in the formation of weak biofilms possibly

Nanosystem	Synthesis process	Chemical precursor	Antibiofilm activity assay %	Receptor protein (genes)	Reference
Cu-NPs	Inert gas condensation	Pure copper	†	†	[74]
ZnO nanostructures from *Nigella sativa* seed	Microwaves	Zinc nitrate and *Nigella sativa* seed extract	91% (256 μg·mL^{-1})	†	[75]
Superparamagnetic IO-NPs	Precipitation and heating at 80°C for 1 h	Ferric chloride and ferric sulfate with polyethylene glycol	88% (16 μg·mL^{-1})	†	[76]
Ag-NPs from grown shoots of *Tamarix nilotica*	Bio-fabrication of Ag-NPs with *T. nilotica*	Silver nitrate with *T. nilotica* extract	62–64% (8 μg·mL^{-1})	†	[77]

†Not declared.

Table 7.
Biofilm formation inhibitor nanosystems tested in vitro *and* in silico *against* L. monocytogenes.

by reducing the surface adhesion and subsequent microcolony formation [75]. Although further research is necessary to unearth the plausible mechanism of biofilm inhibition by the ZnO nanoparticles.

In superparamagnetic iron oxide (IO) nanoparticles, biofilm reduction is attributed to the generation of reactive oxygen species (ROS) due to the interactions of the nanoparticles in the microorganism [76]. Similarly, as shown by Al-Shabib [77], green synthesis of silver nanoparticles produces ROS, causing cell death and inhibition of biofilms. Also, enhanced ROS production as the plausible mechanism of antibiofilm action is described [78] since gold nanoparticles interfere with the EPS matrix and disintegrate the architecture of the biofilm.

7. *Clostridium*

Clostridium is a genus of bacteria that include more than 100 species, is categorized as a Gram-positive bacterium, has flagella, and is anaerobic. They can cause multiple foodborne diseases in humans, such as botulism, *C. perfringens* food poisoning, necrotizing enteritis, and others [5]. These microorganisms can sporulate, which increases their resistance, and they can spread through abiotic surfaces.

In addition, some species can form biofilms, which are beneficial in some industrial processes (recycling and cellulose degradation processes, gas, acetone, butanol, and ethanol production) [5, 79, 80]. There are also some non-pathogenic bacteria in humans that form this type of biofilm and are part of the intestinal microbiota, such as *Clostridium clostridioforme* and *Clostridium malenominatum* [5].

In the case of pathogenic bacteria that cause diseases in humans, biofilms produced from bacteria can be a significant issue because this mechanism protects the microorganisms from antibiotics, the environment, toxic molecules, certain stress conditions, and immune system responses, hence, biofilm helps bacteria survive and may play a role in virulence [5, 6].

Specifically, the *Clostridium* species transmitted by food contamination that causes virulence in humans with the ability to form biofilms are *C difficile*, *C. botulinum,* and *C. perfringens* [5].

C difficile. *Clostridium difficile* infections (CDI) from *C. difficile* are the most prevalent cause of nosocomial diarrhea and colitis in the United States [81, 82]. This species develops in the colon after antibiotic medication changes the gut microbiota and secretes toxins that are virulence factors, including A and B toxins [80, 81, 83].

C. botulinum. These species produce one of the most lethal substances known to induce botulism, botulinum neurotoxin (BoNT). Botulinum toxin inhibits nerve function and can induce paralysis of the respiratory and muscular systems [84].

C. perfringens. One of the most common causes of foodborne infection in the United States. It can produce numerous toxins and is the cause of gas gangrene, necrotizing enteritis, food poisoning, and diarrhea associated with antibiotics [6].

7.1 Quorum sensing related genes and nanosystems' evaluation

Into the *Clostridium* genus, the implicated QS genes of *C. perfringens* are well established. QS is performed by the Agr and LuxS systems that are involved in toxin production and pathogenicity through propeptides and AI-2 production, respectively [85]. Although CDI have been widely reported and associated with *C. difficile* toxin production, the QS system by which these infections are regulated is considered a complex multifactorial process [86]. There are even few reports on the effect of nanosystems involved in biofilm and QS inhibition in *Clostridium.*

In a study, Omoigberale, see [87], it has been found that Ag-NPs at a concentration of 10 mg·mL^{-1} can generate a biofilm reduction of 19–58%, while Au-NPs generate a smaller reduction with 12–39%, at the same concentration. However, this percentage range of reduction is considered strain-dependent, since this behavior was evaluated in 17 strains of *Clostridia.* At the same time, Au-NPs/Ag-NPs-Gentamicin complexes were evaluated obtaining 31% and 30% biofilm reduction, respectively. Although it was not a better performance than that obtained for the simple metallic nanoparticles, it did have a better effect compared to other antibiotics tested in the nanosystems.

Silver nanoparticles (Ag-NPs) also have been evaluated on specific isolates of *C. perfringens*, finding biofilm inhibition percentages of 80.8–82.8% at concentrations up to 100 µg·mL^{-1}, but showing the same behavior as in previous studies [87] in which these percentages are strain-dependent, or in this case, isolate-dependent from different animal and human organisms [88].

Undoubtedly, the molecular study of nanosystems with inhibitory QS potential is a great area of opportunity for this genus, pursuing to elucidate the signaling mechanisms and QQs that could counteract its pathogenicity.

8. Conclusions and perspectives

Biofilm formation is nowadays a worldwide topic of interest for public health due to the importance in pathogen survival. Knowing the mechanism of synthesis can help to develop new materials to prevent and eliminate bacteria. Quorum sensing is one of the mechanisms through which bacteria can communicate and create barriers against antibacterial agents. Nanosystems applied in food safety may interact with genes that express the signalization of quorum sensing, as many authors have reported. Understanding the interaction of the nanoparticle with the DNA of cells may lead to the formulation of new materials that exhibit this inhibition route, and for that, molecular docking can help to elucidate the possible interaction and, further, allow the food industry to be even safer for all populations.

Acknowledgements

The authors express their gratitude to the Mexican National Council of Science and Technology (CONACYT) for their support under the grant number 973558 throughout the Doctoral program in Sciences in Biotechnological Processes of the Universidad de Guadalajara.

Conflict of interest

The authors declare no conflict of interest.

Notes/thanks/other declarations

The authors state that there are no other declarations.

Author details

Dulce María Romero-García[1], Jazmín Guadalupe Silva-Galindo[1], Carlos Arnulfo Velázquez-Carriles[2], Alejandra Monserrat Castorena-Sánchez[1], Ismael García-Vera[1], Gabriela Hinojosa-Ventura[1] and Jorge Manuel Silva-Jara[1]*

1 University Center of Exact Sciences and Engineering (CUCEI), Universidad de Guadalajara, Guadalajara, Mexico

2 University Center of Tlajomulco (CUTLAJO), Universidad de Guadalajara, Guadalajara, Mexico

*Address all correspondence to: jorge.silva@academicos.udg.mx

References

[1] Galié S, García-Gutiérrez C, Miguélez EM, Villar CJ, Lombó F. Biofilms in the food industry: Health aspects and control methods. Frontiers in Microbiology. 2018;**9**

[2] Zhao X, Zhao F, Wang J, Zhong N. Biofilm formation and control strategies of foodborne pathogens: Food safety perspectives. RSC Advances. 2017;7(58): 36670-36683

[3] Liu X, Yao H, Zhao X, Ge C. Biofilm formation and control of foodborne pathogenic bacteria. Molecules. 2023; **28**(6):2432

[4] Kirmusaoglu S. Staphylococcal biofilms: Pathogenicity, mechanism and regulation of biofilm formation by quorum-sensing system and antibiotic resistance mechanisms of biofilm-embedded microorganisms. In: Microbial Biofilms - Importance and Applications. London: InTech; 2016

[5] Pantaléon V, Bouttier S, Soavelomandroso AP, Janoir C, Candela T. Biofilms of *Clostridium* species. Anaerobe. 2014;**30**:193-198

[6] Charlebois A, Jacques M, Archambault M. Comparative transcriptomic analysis of *Clostridium perfringens* biofilms and planktonic cells. Avian Pathology. 2016;**45**(5): 593-601

[7] Kim SR, Yeon KM. Quorum Sensing as Language of Chemical Signals. Comprehensive Analytical Chemistry. 2018;**81**:57-94

[8] Vogel J, Quax WJ. Enzymatic Quorum Quenching in Biofilms. Quorum Sensing. Molecular Mechanism and Biotechnological Application. 2019;**7**: 173-193

[9] Bai X, Nakatsu CH, Bhunia AK. Bacterial biofilms and their implications in pathogenesis and food safety. Food. 2021;**10**(9):2117

[10] Bu F, Liu M, Xie Z, Chen X, Li G, Wang X. Targeted anti-biofilm therapy: Dissecting targets in the biofilm life cycle. Pharmaceuticals. 2022;**15**(10): 1253

[11] Martínez A, Stashenko EE, Sáez RT, Zafra G, Ortiz C. Effect of essential oil from Lippia origanoides on the transcriptional expression of genes related to quorum sensing, biofilm formation, and virulence of *Escherichia coli* and *Staphylococcus aureus*. Antibiotics. 2023;**12**(5):845

[12] Zhang X, Liu B, Ding X, Bin P, Yang Y, Zhu G. Regulatory mechanisms between quorum sensing and virulence in *Salmonella*. Microorganisms. 2022; **10**(11):2211

[13] Khadke SK, Lee J-H, Kim Y-G, Raj V, Lee J. Appraisal of cinnamaldehyde analogs as dual-acting antibiofilm and anthelmintic agents. Frontiers in Microbiology. 2022;**13**:1-16

[14] Ren Q, Luo W, Chi H, Zhang L, Chen W. Down-regulation of β-lactam antibiotics resistance and biofilm formation by *Staphylococcus epidermidis* is associated with isookanin. Frontiers in Cellular and Infection Microbiology. 2023;**13**

[15] Sung K, Khan S, Ahn J. Foodborne pathogen biofilms: Development, detection, control, and antimicrobial resistance. Pathogens. 2023;**12**(2):352

[16] Hall CW, Mah TF. Molecular mechanisms of biofilm-based antibiotic resistance and tolerance in pathogenic

bacteria. FEMS Microbiology Reviews. 2017;**41**(3):276-301

[17] Soheili V, Bazzaz BSF, Abdollahpour N, Hadizadeh F. Investigation of *Pseudomonas aeruginosa* quorum-sensing signaling system for identifying multiple inhibitors using molecular docking and structural analysis methodology. Microbial Pathogenesis. 2015;**89**:73-78

[18] Sauer K, Camper AK, Ehrlich GD, Costerton JW, Davies DG. *Pseudomonas aeruginosa* displays multiple phenotypes during development as a biofilm. Journal of Bacteriology. 2002;**184**(4):1140-1154

[19] Jeong GJ, Khan F, Khan S, Tabassum N, Mehta S, Kim YM. *Pseudomonas aeruginosa* virulence attenuation by inhibiting siderophore functions. Applied Microbiology and Biotechnology. 2023;**107**(4):1019-1038

[20] dos Siqueira FS, dos Alves CFS, Machado AK, Siqueira JD, dos Santos T, Mizdal CR, et al. Molecular docking, quorum quenching effect, antibiofilm activity and safety profile of silver-complexed sulfonamide on *Pseudomonas aeruginosa*. Biofouling. 2021;**37**(5):555-571

[21] MohdF K, Husain FM, Zia Q, Ahmad E, Jamal A, Alaidarous M, et al. Anti-quorum sensing and anti-biofilm activity of zinc oxide nanospikes. ACS Omega. 2020;**5**(50):32203-32215

[22] Prateeksha SBR, Shoeb M, Sharma S, Naqvi AH, Gupta VK, et al. Scaffold of selenium nanovectors and honey phytochemicals for inhibition of *Pseudomonas aeruginosa* quorum sensing and biofilm formation. Frontiers in Cellular and Infection Microbiology. 2017;**7**:93

[23] Shah S, Gaikwad S, Nagar S, Kulshrestha S, Vaidya V, Nawani N, et al. Biofilm inhibition and anti-quorum sensing activity of phytosynthesized silver nanoparticles against the nosocomial pathogen *Pseudomonas aeruginosa*. Biofouling. 2019;**35**(1): 34-49

[24] Elshaer SL, Shaaban MI. Inhibition of quorum sensing and virulence factors of *Pseudomonas aeruginosa* by biologically synthesized gold and selenium nanoparticles. Antibiotics. 2021;**10**(12): 1461

[25] O'Toole GA, Kolter R. Initiation of biofilm formation in *Pseudomonas fluorescens* WCS365 proceeds via multiple, convergent signalling pathways: a genetic analysis. Molecular Microbiology. 1998;**28**(3):449-461

[26] Sánchez-Vargas FM, Abu-El-Haija MA, Gómez-Duarte OG. *Salmonella* infections: An update on epidemiology, management, and prevention. Travel Medicine and Infectious Disease. 2011;**9**(6):263-277

[27] González JF, Alberts H, Lee J, Doolittle L, Gunn JS. Biofilm formation protects *salmonella* from the antibiotic ciprofloxacin *in vitro* and *in vivo* in the mouse model of chronic carriage. Scientific Reports. 2018;**8**(1):222

[28] Gong B, Li H, Feng Y, Zeng S, Zhuo Z, Luo J, et al. Prevalence, serotype distribution and antimicrobial resistance of non-typhoidal *salmonella* in hospitalized patients in Conghua district of Guangzhou, China. Frontiers in Cellular and Infection Microbiology. 2022;**12**:805384

[29] Torgerson PR, Devleesschauwer B, Praet N, Speybroeck N, Willingham AL, Kasuga F, et al. World Health organization estimates of the global and regional disease burden of 11 foodborne

parasitic diseases, 2010: A data synthesis. PLoS Medicine. 2015;**12**(12):e1001920

[30] Ferrari RG, Rosario DKA, Cunha-Neto A, Mano SB, Figueiredo EES, Conte-Junior CA. Worldwide epidemiology of *salmonella* serovars in animal-based foods: A meta-analysis. Applied and Environmental Microbiology. 2019; **85**(14):e00591-19

[31] He Y, Wang J, Zhang R, Chen L, Zhang H, Qi X, et al. Epidemiology of foodborne diseases caused by *Salmonella* in Zhejiang Province, China, between 2010 and 2021. Frontiers in Public Health. 2023;**11**:1127925

[32] Popa GL, Popa MI. *Salmonella* spp. infection – a continuous threat worldwide. Germs. 2021;**11**(1):88-96

[33] Preda VG, Săndulescu O. Communication is the key: biofilms, quorum sensing, formation and prevention. Discover. 2019; **7**(3):e10

[34] Jahan F, Chinni SV, Samuggam S, Reddy LV, Solayappan M, Su YL. The complex mechanism of the *Salmonella typhi* biofilm formation that facilitates pathogenicity: A review. International Journal of Molecular Sciences. 2022; **23**(12):6462

[35] Sholpan A, Lamas A, Cepeda A, Franco CM. *Salmonella* spp. quorum sensing: an overview from environmental persistence to host cell invasion. AIMS Microbiology. 2021;**7**(2): 238-256

[36] Sheng X, Wang W, Chen L, Zhang H, Zhang Y, Xu S, et al. Mig-14 may contribute to *Salmonella enterica* serovar typhi resistance to polymyxin B by decreasing the permeability of the outer-membrane and promoting the formation of biofilm. International Journal of Medical Microbiology. 2019; **309**(2):143-150

[37] Chen L, Gu L, Geng X, Xu G, Huang X, Zhu X. A novel cis antisense RNA AsfD promotes *Salmonella enterica* serovar typhi motility and biofilm formation. Microbial Pathogenesis. 2020;**142**:104044

[38] de Almeida FA, Vargas ELG, Carneiro DG, Pinto UM, Vanetti MCD. Virtual screening of plant compounds and nonsteroidal anti-inflammatory drugs for inhibition of quorum sensing and biofilm formation in *Salmonella*. Microbial Pathogenesis. 2018;**121**: 369-388

[39] Olaniyi TD, Adetutu A. In silico anti-quorum sensing activities of phytocompounds of Psidium guajava in *Salmonella enterica* serovar typhi. Journal of Umm Al-Qura University for Applied Sciences. 2023;**9**:142-156

[40] Aswathanarayan JB, Vittal RR. Inhibition of biofilm formation and quorum sensing mediated phenotypes by berberine in *Pseudomonas aeruginosa* and *Salmonella typhimurium*. RSC Advances. 2018;**8**(63):36133-36141

[41] Almasoud A, Hettiarachchy N, Rayaprolu S, Babu D, Kwon YM, Mauromoustakos A. Inhibitory effects of lactic and malic organic acids on autoinducer type 2 (AI-2) quorum sensing of *Escherichia coli* O157:H7 and *Salmonella* typhimurium. LWT - Food Science and Technology. 2016;**66**: 560-564

[42] Balakrishnan S, Ibrahim KS, Duraisamy S, Sivaji I, Kandasamy S, Kumarasamy A, et al. Antiquorum sensing and antibiofilm potential of biosynthesized silver nanoparticles of

Myristica fragrans seed extract against MDR *Salmonella enterica* serovar Typhi isolates from asymptomatic typhoid carriers and typhoid patients. Environmental Science and Pollution Research. 2020;**27**(3):2844-2856

[43] Xin W-G, Wu G, Ying J-P, Xiang Y-Z, Jiang Y-H, Deng X-Y, et al. Antibacterial activity and mechanism of action of bacteriocin LFX01 against *Staphylococcus aureus* and *Escherichia coli* and its application on pork model. Meat Science. 2023;**196**:109045

[44] Styles MJ, Early SA, Tucholski T, West KHJ, Ge Y, Blackwell HE. Chemical control of quorum sensing in *E. coli*: Identification of small molecule modulators of SdiA and mechanistic characterization of a covalent inhibitor. ACS Infectious Diseases. 2020;**6**(12): 3092-3103

[45] Vila-Sanjurjo C, Hembach L, Netzer J, Remuñán-López C, Vila-Sanjurjo A, Goycoolea FM. Covalently and ionically, dually crosslinked chitosan nanoparticles block quorum sensing and affect bacterial cell growth on a cell-density dependent manner. Journal of Colloid and Interface Science. 2020;**578**: 171-183

[46] Qin X, Kräft T, Goycoolea FM. Chitosan encapsulation modulates the effect of trans -cinnamaldehyde on AHL-regulated quorum sensing activity. Colloids Surfaces B Biointerfaces. 2018; **169**:453-461

[47] Thanh Nguyen H, Goycoolea F. Chitosan/cyclodextrin/TPP nanoparticles loaded with quercetin as novel bacterial quorum sensing inhibitors. Molecules. 2017;**22**(11):1975

[48] Wang R, Vega P, Xu Y, Chen C-Y, Irudayaraj J. Exploring the anti-quorum sensing activity of a d−limonene nanoemulsion for *Escherichia coli* O157: H7. Journal of Biomedial Materials Research Part A. 2018;**106**(7):1979-1986

[49] Ejaz S, Ihsan A, Noor T, Shabbir S, Imran M. Mannose functionalized chitosan nanosystems for enhanced antimicrobial activity against multidrug resistant pathogens. Polymer Testing. 2020;**91**:106814

[50] Zelelew D, Endale M, Melaku Y, Demissie TB, Ombito JO, Eswaramoorthy R. ZnO nanoparticle-assisted synthesis of Thiazolo[3,2-α] Pyrimidine analogs: Antibacterial and antioxidant activity, *in silico* molecular docking, and ADMET prediction study. Journal of Chemistry. 2022;**2022**: 1-20

[51] Schembri MA, Kjaergaard K, Klemm P. Global gene expression in *Escherichia coli* biofilms. Molecular Microbiology. 2003;**48**(1):253-267

[52] Álvarez-Chimal R, García-Pérez VI, Álvarez-Pérez MA, Tavera-Hernández R, Reyes-Carmona L, Martínez-Hernández M, et al. Influence of the particle size on the antibacterial activity of green synthesized zinc oxide nanoparticles using *Dysphania ambrosioides* extract, supported by molecular docking analysis. Arabian Journal of Chemistry. 2022;**15**(6):103804

[53] Alarfaj AA, Hajinur Hirad A, Ravindran B, Perumal NS. Evaluation of anti-microbial activity and in-silico docking and Adme studies of green synthesised copper oxide nanoparticle from *Aegle marmelos* leaf extract. SSRN. 2023;**31**

[54] Riedel S, Hobden JA, Miller S, Morse SA, Mietzner TA, Detrick B, et al. Jawetz, Melnick and Adelberg's Medical

Microbiology. 28th edition. McGraw Hill; 2019. ISBN 978-1-260-01202-6

[55] Kourtis AP, Hatfield K, Baggs J, Mu Y, See I, Epson E, et al. *Vital signs:* epidemiology and recent trends in methicillin-resistant and in methicillin-susceptible *Staphylococcus aureus* bloodstream infections — United States. MMWR. Morbidity and Mortality Weekly Report. 2019;**68**(9):214-219

[56] Turner NA, Sharma-Kuinkel BK, Maskarinec SA, Eichenberger EM, Shah PP, Carugati M, et al. Methicillin-resistant *Staphylococcus aureus*: an overview of basic and clinical research. Nature Reviews. Microbiology. 2019; **17**(4):203-218

[57] Tan L, Li SR, Jiang B, Hu XM, Li S. Therapeutic targeting of the *Staphylococcus aureus* accessory gene regulator (agr) system. Frontiers in Microbiology. 2018;**9**:55

[58] Dehbashi S, Tahmasebi H, Zeyni B, Arabestani MR. The relationship between promoter-dependent quorum sensing induced genes and methicillin resistance in clinical strains of *Staphylococcus aureus*. Journal of Advances in Medical and Biomedical Research. 2018;**26**(116):75-87

[59] Parastan R, Kargar M, Solhjoo K, Kafilzadeh F. *Staphylococcus aureus* biofilms: Structures, antibiotic resistance, inhibition, and vaccines. Gene Reports. 2020;**20**:100739

[60] Chen Q, Xie S, Lou X, Cheng S, Liu X, Zheng W, et al. Biofilm formation and prevalence of adhesion genes among *Staphylococcus aureus* isolates from different food sources. Microbiology. 2020;**9**(1):e00946

[61] Piewngam P, Chiou J, Chatterjee P, Otto M. Alternative approaches to treat bacterial infections: Targeting quorum-sensing. Expert Review of Anti-Infective Therapy. 2020;**18**(6):499-510

[62] Rashki S, Ghanbari M, Khudhair ZH, Marzhoseyni Z, Bameri Z, Afsharikhah S, et al. Evaluate the effect of graphitic carbon nitride nanosheets decorated with copper nanoparticles on biofilm and icaA gene expression in *Staphylococcus aureus* isolated from clinical samples. Arabian Journal of Chemistry. 2023;**16**(8):104882

[63] Ramachandran G, Alharbi NS, Chackaravarthy G, Kanisha Chelliah C, Rajivgandhi G, Maruthupandy M, et al. Chitosan/silver nanocomposites enhanced the biofilm eradication in biofilm forming Gram positive *S. aureus*. Journal of King Saud University - Science. 2023;**35**(4):102597

[64] Muthuchamy M, Govindan R, Shine K, Thangasamy V, Alharbi NS, Thillaichidambaram M, et al. Anti-biofilm investigation of graphene/chitosan nanocomposites against biofilm producing *P. aeruginosa* and *K. pneumoniae*. Carbohydrate Polymers. 2020;**230**:115646

[65] Hemmati F, Salehi R, Ghotaslou R, Kafil HS, Hasani A, Gholizadeh P, et al. The assessment of antibiofilm activity of chitosan-zinc oxide-gentamicin nanocomposite on *Pseudomonas aeruginosa* and *Staphylococcus aureus*. International Journal of Biological Macromolecules. 2020;**163**:2248-2258

[66] Centers for Disease Control and Prevention (CDC). *Listeria* (Listeriosis). Available from: https://www.cdc.gov/spanish/listeria/index.html. 2017

[67] Shamloo E, Hosseini H, Moghadam ZA, Larsen MH, Haslberger A, Alebouyeh M. Importance of *Listeria monocytogenes* in food safety: A

review of its prevalence, detection, and antibiotic resistance. Iranian Journal of Veterinary Research. 2019;**20**(4):241

[68] Lopes-Luz L, Mendonça M, Bernardes Fogaça M, Kipnis A, Bhunia AK, Bührer-Sékula S. *Listeria monocytogenes*: Review of pathogenesis and virulence determinants-targeted immunological assays. Critical Reviews in Microbiology. 2021;**47**(5):647-666

[69] Osek J, Lachtara B, Wieczorek K. *Listeria monocytogenes* in foods—From culture identification to whole-genome characteristics. Food Science & Nutrition. 2022;**10**(9):2825-2854

[70] Jamshidi A, Zeinali T. Significance and Characteristics of *Listeria monocytogenes* in Poultry Products. International Journal of Food Science. 2019;**2019**:1-7

[71] Gemmell CT, Parreira VR, Farber JM. Controlling *Listeria monocytogenes* growth and biofilm formation using flavonoids. Journal of Food Protection. 2022;**85**(4): 639-646

[72] Guo L, Ren C, Chen G, Wu M, Liu W, Ding C, et al. Reactive oxygen species inhibit biofilm formation of *Listeria monocytogenes*. Microbial Pathogenesis. 2019;**127**:183-189

[73] Hanes RM, Huang Z. Investigation of antimicrobial resistance genes in *Listeria monocytogenes* from 2010 through to 2021. International Journal of Environmental Research and Public Health. 2022;**19**(9):5506

[74] Ghasemian E, Naghoni A, Rahvar H, Kialha M, Tabaraie B. Evaluating the effect of copper nanoparticles in inhibiting *Pseudomonas aeruginosa* and *Listeria monocytogenes* biofilm formation. Jundishapur. Journal of Microbiology. 2015;**8**(5):e17430

[75] Al-Shabib NA, Husain FM, Ahmed F, Khan RA, Ahmad I, Alsharaeh E, et al. Biogenic synthesis of zinc oxide nanostructures from *Nigella sativa* seed: Prospective role as food packaging material inhibiting broad-spectrum quorum sensing and biofilm. Scientific Reports. 2016;**6**(1):36761

[76] Al-Shabib NA, Husain FM, Ahmed F, Khan RA, Khan MS, Ansari FA, et al. Low temperature synthesis of superparamagnetic iron oxide (Fe_3O_4) nanoparticles and their ROS mediated inhibition of biofilm formed by food-associated bacteria. Frontiers in Microbiology. 2018;**9**:2567

[77] Al-Shabib NA, Husain FM, Nadeem M, Khan MS, Al-Qurainy F, Alyousef AA, et al. Bio-inspired facile fabrication of silver nanoparticles from *in vitro* grown shoots of *Tamarix nilotica*: explication of its potential in impeding growth and biofilms of *Listeria monocytogenes* and assessment of wound healing ability. RSC Advances. 2020; **10**(50):30139-30149

[78] Perveen K, Husain FM, Qais FA, Khan A, Razak S, Afsar T, et al. Microwave-assisted rapid green synthesis of gold nanoparticles using seed extract of *Trachyspermum ammi*: ROS mediated biofilm inhibition and anticancer activity. Biomolecules. 2021; **11**(2):197

[79] Dumitrache A, Wolfaardt GM, Allen DG, Liss SN, Lynd LR. Tracking the cellulolytic activity of *Clostridium thermocellum* biofilms. Biotechnology for Biofuels. 2013;**6**(1):175

[80] Engevik MA, Luk B, Hall A, Ganesh BP, Versalovic J. Mo1658 *clostridium difficile* chemotaxes toward

mucin glycans, adheres and forms biofilms *in vitro* and *in vivo*. Gastroenterology. 2016;**150**(4):S745

[81] Darkoh C, DuPont HL, Norris SJ, Kaplan HB. Toxin synthesis by *Clostridium difficile* is regulated through quorum signaling. MBio. 2015;**6**(2): e02569

[82] James GA, Chesnel L, Boegli L, deLancey PE, Fisher S, Stewart PS. Analysis of *Clostridium difficile* biofilms: Imaging and antimicrobial treatment. Journal of Antimicrobial Chemotherapy. 2018;**73**(1):102-108

[83] Murase T, Eugenio L, Schorr M, Hussack G, Tanha J, Kitova EN, et al. Structural basis for antibody recognition in the receptor-binding domains of toxins a and b from *clostridium difficile*. Journal of Biological Chemistry. 2014; **289**(4):2331-2343

[84] Halpin JL, Dykes JK, Katz L, Centurioni DA, Perry MJ, Egan CT, et al. Molecular characterization of *clostridium botulinum* harboring the *bont* /B7 Gene. Foodborne Pathogens and Disease. 2019; **16**(6):428-433

[85] Xu Y, Wang Y, Ding X, Wang J, Zhan X. Inhibitory effects of reuterin on biofilm formation, quorum sensing and virulence genes of *Clostridium perfringens*. LWT. 2022;**162**:113421

[86] Thapa T, Leuzzi R, Ng YK, Baban ST, Adamo R, Kuehne SA, et al. Multiple factors modulate biofilm formation by the anaerobic pathogen *Clostridium difficile*. Journal of Bacteriology. 2013;**195**(3):545-555

[87] Omoigberale MO. Evaluating the impact of alternative antimicrobials on biofilms formed by *Clostridium perfringens*. [Doctoral Thesis]: Lincoln, University of Lincoln; 2021

[88] Ahmed HA, El Bayomi RM, Hamed RI, Mohsen RA, El-Gohary FA, Hefny AA, et al. Genetic relatedness, antibiotic resistance, and effect of silver nanoparticle on biofilm formation by *Clostridium perfringens* isolated from chickens, pigeons, camels, and human consumers. Veterinary Sciences. 2022; **9**(3):109

Chapter 8

Efficacy of Natural and Synthetic Biofilm Inhibitors Associated with Antibiotics in Eradicating Biofilms Formed by Multidrug-Resistant Bacteria

Salma Kloula Ben Ghorbal, Sana Dhaouadi, Sana Bouzenbila, Ameur Cherif and Ramzi Boubaker Elandoulsi

Abstract

Biofilms formed by multidrug resistant (MDR) bacteria like methicillin-resistant *Staphylococcus aureus* (MRSA) and others are the main causes of infections that represent a serious public health issue. Persistent MDR infections are mostly derived from biofilm formation which in turn leads to resistance to conventional antimicrobial therapy. Inhibition of bacterial surface attachment is the new alternative strategy without affecting the bacterial growth. Thus, the discovery of compounds that interfere with biofilm production, virulence factors release and quorum sensing (QS) detection in pathogens is a promising processus. Among these compounds, natural and synthetic molecules are a compelling alternative to attenuate pathogenicity. The combination of these compounds with antibiotics makes the bacteria more vulnerable to the later, once used alone. This combination can restore antibiotic effectiveness against MDR bacteria. Among these molecules, 3-phenylpropan-1-amine (3-PPA) has been found to inhibit *Serratia marcescens* biofilm formation, PAβN has been proven to inhibit biofilm prodcution in *A. baumannii,* while brominated Furanone C-30 has been reported to be a potent inhibitor of the QS system and *P. aeruginosa* biofilm. Therefore, the combination between biofilm-inhibitors and antibiotics represents a promising strategy to mitigate antibiotic resistance in MDR pathogens, which has become a major threat to public healthcare around the globe.

Keywords: biofilm inhibitors, antibiotics, association, MDR bacteria, biofilm

1. Introduction

Diseases that are caused by pathogens producing bacterial biofilms are increasingly spread, which represent a real threat to human health. Therefore, floating or swimming bacteria are more vulnerable to antibiotics. However, they can be reorganized into clusters of very complex structure, composed of a matrix of self-synthesized exopolymers, which forms the notorious biofilm that is hard to eradicate because bacteria embedded in this structure become highly resistant to many antimicrobial agents. In fact, when trapped in biofilms, biofilm-producing bacteria can be over 1000-fold more resistant to antimicrobials than their planktonic equivalents [1]. In addition, the massive use of antimicrobials has led to an increment in multi-drug resistance (MDR) of pathogenic bacteria, rendering the failure of antibiotic treatment. The six main multidrug-resistant and fatal pathogens are known as "ESKAPE" pathogens: *Escherichia coli*, *S. aureus*, *Klebsiella pneumoniae*, *A. baumannii*, *P. aeruginosa and Enterobacter* spp. These bacterial agents are responsible for polymicrobial infections that cause diseases such as cystic fibrosis, ear and urinary tract infections, respiratory tract infections, diabetic ulcers, wounds, in addition to the contamination of certain medical devices [2]. Furthermore, majority of chronic and nosocomial infections are associated with mono- or polymicrobial biofilms, having a significant impact on the survival rates of patients. Although the use of medical devices revolutionized health care services and significantly improved patient outcomes, it also led to complications due to the associations with biofilms and the emergence of multidrug resistant bacteria.

In particular, MDR bacteria poses a major challenge as current antimicrobial therapies are often associated with poor outcomes [3]. Based on the progress of the mechanism of biofilm development in MDR bacteria, many anti-biofilm molecules are being discovered with diverse modes of action such as quorum quenching (QQ) and cell adhesion inhibition, dispersion of extracellular polymeric substance, and interference with c-di-GMP signaling pathways, etc. [4]. Taking these factors into account, it is clear that new strategies are required to weaken the biofilm, inhibiting its proliferation and making it less resistant to antibiotics. These strategies involve targetting the resistance mechanisms of pathogenic bacteria like the production of biofilms by controlling quorum sensing (QS) since it is an intercellular communication system, which influences microbial virulence [5]. Therefore, interference in QS system of bacterial pathogens can reduce drug resistance which is considered as a suitable alternative that attenuates pathogenicity and protects the host from infection due to biofilm formation [1].

This issue has prompted researchers to find new microbial biofilm inhibitors that could be combined with existing antibiotics to improve their efficacy in bacterial eradication. In recent years, researchers have increasingly sought alternative therapeutic strategies for effective treatment of biofilm-producing pathogens. The target is to overcome the drawbacks of conventional antimicrobial therapies as microbial infections involving biofilms become quite challenging because of their high antibiotic resistance capacities. Within this framework, the present study has evaluated the anti-biofilm characteristics of natural and synthetic molecules against MDR bacteria.

2. Anti-biofilm activity of peptides and organic compounds

2.1 Anti-biofilm activity of 3-Phenylpropan-1-amine (3-PPA)

Phenylpropane-1-amine (3-PPA) is known to be an antibiotic adjuvant that interferes with QS and disrupts signaling between bacteria without posing a threat to the bacteria themselves, potentially resolving resistance in pathogenic bacteria [6]. In this recent and unique study, 3-phenylpropan-1-amine (3-PPA) was determined to inhibit biofilm formation. Furthermore, the inhibitory effect rises along with high drug concentrations. Notably, at 50 μg/mL, 3-PPA treatment reduces biofilm formation by 48%. Moreover, 3-PPA probably acts on virulence factors. They also studied the expression of genes related to detoxification enzymes and found a 37% inhibition in the expression of *sodB* gene, which encodes superoxide dismutase (SOD). Given the inhibitory effects of 3-PPA on biofilm formation, they also explored whether 3-PPA can increase the vulnerability of biofilms to traditional antibiotics. Thus, biofilms were exposed to 3-PPA and antibiotics in combination. In fact, 3-PPA (50 μg/mL) or ofloxacin (0.2 μg/mL) alone had weak effects on biofilm eradication, but relatively strong effects when used in combination, with a biofilm erasure rate of 44%. They also confirmed that by scanning electron microscopy (SEM), treatment with the combination of 3-PPA and ofloxacin resulted in the significant dispersal, destruction, and reduction of the preformed biofilm. Therefore, 3-PPA was used as an antibiotic adjuvant to interfere with the QS and interrupt the signaling between bacteria while not being a threat to the microorganism, which could help solve the problem of resistance in disease-causing bacteria. This is the only work to develop a strategy to by-pass multidrug-resistant *S. marcescens* and improve treatment outcomes for recalcitrant infections (**Table 1**).

2.2 Anti-biofilm activity of cathelicidin-related antimicrobial peptide (CRAMP)

De Brucker et al. [7] identified AS10 (Peptide Sequene: KLKKIAQKIKNFFQKLVP) as the most potent anti-biofilm peptide at 0.22 M. This peptide inhibits biofilm formation of the fungus *C. albicans* and also various Gram-positive and Gram-negative bacteria in a mixed biofilm and acts synergistically with caspofungin or amphotericin B against mature *C. albicans* biofilm. Recently, in the study by Zhang et al. [8], the best synergistic activity of CRAMP combined with colistin at 62.5 μg/ml was confirmed for *P. aeruginosa*, with a significant inhibition of the biomass of preformed biofilms reaching 91.05%, confirmed by confocal laser scanning microscopy (CLSM) images. It was also confirmed that the combination (CRAMP-1/4 MIC colistin) also down-regulated the expression of QS regulated genes, including pyocyanin and rhamnolipid production [9]. In 2022, the same research team also elucidated the specific mechanism by which CRAMP was able to eradicate *P. aeruginosa* biofilms using an integrative analysis of transcriptomic, proteomic and metabolomic data [10]. Somal data showed that CRAMP acts on *P. aeruginosa* biofilms through a range of pathways, which include the *Pseudomonas* quinolone signaling system (PQS), the cyclic dimeric guanosine monophosphate (c-di-GMP) signaling pathway, and the exopolysaccharide and rhamnolipid synthesizing pathways [10]. These studies provide new possibilities for the development of CRAMP as a potentially effective anti-biofilm dispersant or even a biofilm-preventive coating for implants (**Table 1**).

2.3 Anti-staphyloxantin activity of NP16 and Celastrol in *S. aureus* biofilm

In the recent study by Gao et al. [11], it was demonstrated a novel inhibitor (NP16) of *S. aureus* staphyloxantin (STX) production. This inhibitor targets the dehydrosqualene desaturase (CrtN) which catalyzes the first step of the staphyloxantin biosynthetic pathway. Staphyloxantin inhibition can reduce the survival of *S. aureus* under oxidative stress conditions and limits biofilm formation. This newly discovered CrtN inhibitor NP16 may represent an effective strategy for combating *S. aureus* biofilms. This molecule is not the only one to have an anti-

Biofilm Inhibitors	Composition or structure	Dose	MDR	Target	Associated antibiotic	References
3-phenylpropan-1-amine (3-PPA)	Peptide	50 mg/mL	*Serratia marescens* NJ01	1. Biofilm inhibition (48%) 2. Virulence factors: i. Prodigiosin ii. Protease iii. Lipase iv. Superoxide dismutase v. Hemolysine vi. Swimming	1. PPA (50 μ g/ml) + Ofloxacine (0.2 μ g/ml)	[6]
AS10	Peptide	0.22 M	*Candida albicans* and various Gram-positive and Gram-negative bacteria Mixed biofilm	Biofilm inhibition	AS10 (0.22 M) +0 Caspofungin or Amphotericin B	[7]
CRAMP	peptide	NM	*P. aeruginosa*	1. 91.05% of biofilm inhibition 2. QS regulated genes: i. Pyocyanin ii. Rhamnolipid production	CRAMP + Colistin (1/4 MIC)	[8, 10]
NP16	Peptide	NM	Methicillin resistant *S. aureus* (MRSA)	1. Antistaphyloxantin production 2. Dehydrosqualene desaturase (CrtN) 3. Anti-biofilm 4. Expolysacharides reduction 5. Membrane integrity	NA	[11]
Celastrol	Organic compound: Pentacyclic triterpenoid	1 µg/mL			Membrane targeting antibiotics: Polymixin B	[12]
Meta-bromo-thiolactone (mBTL)	Synthetic molecule	NM	*P. aeruginosa*	1. 1.Quorum-sensing receptors, LasR and RhlR 2. RhiR is the most relevant target 3. genes encoding pyocyanin	NA	[13]

DOI: http://dx.doi.org/10.5772/intechopen.112408

Biofilm Inhibitors	Composition or structure	Dose	MDR	Target	Associated antibiotic	References
Malondialdehyde (MDA)	Natural/Synthetic molecule	90 µg/ml 180 µg/ml	*Staphylococcus xylosus* *Lactiplantibacillus. plantarum*	1. LDH release 2. Ca^{2+} and Mg^{2+} leakage 3. ATP reduction	NA	[14]
Psammaplin A	Marine-derived bromotyrosine compounds	NM	*P. aeruginosa* PAO1 lasB-gfp *P. aeruginosa* PAO1 rhlA-gfp	Quorum-sensing inhibitor	IC50 value at 30.69 for PAO1 lasB-gfp and IC50 value at 2.64 µM for PAO1 rhlA-gfp	[15]
Bisaprasin					IC50 value at 8.70 µM for PAO1 lasB-gfp and IC50 value at 8.53 µM for PAO1 rhlA-gfp	
Maipomycin A (MaiA)	Marine and bacterial-derived products	NM	• Gram negative bacteria • Actinomycete strain • *Kibdelosporangium phytohabitans* XY-R10	Undefined pathway	Colistin	[8]
Piperine	Naturally occurring alkaloid	NM	*K. pneumoniae*	Preformed biofilm (MBEC)	Kanamycine + piperine (MBEC reduced by 8- to 16-fold)	[16]
Thymol (2-Isopropyl-5-methylphenol)	Essential oil: natural volatile monoterpenoid phenol	NM	Methicillin resistant *S. aureus* (MRSA)	Staphyloxanthin biosynthesis (CrtM)	90% of staphyloxanthin inhibition at 100 µg/mL + Polymixin B	[17]
			K. pneumoniae	Biofilm formation (MBIC)	Streptomycin + thymol (MBIC reduced by 16- to 64-fold)	
					Kanamycin+ piperine (MBIC reduced by 4-fold)	
			K. pneumoniae	Preformed biofilm (MBEC)	Streptomycin + thymol (MBEC reduced by 16- to 128-fold)	[16]
					Amikacine+ thymol (MBEC reduced by 4- to 128-fold)	
					Kanamycine + thymol (MBEC reduced by 8- to 256-fold)	
			Methicillin resistant *S. aureus* (MRSA)	Biofilm formation	Rifampicine +thymol (88% MRSA biofilm reduction)	[18]

Abbrevations: NM: Not mentioned; NA: Not associated; QS: Quorum sensing; MBIC: Minimal biofilm inhibition concentration; MBEC: Minimal biofilm eradication concentration.

Table 1.
List of biofilm inhibitors.

staphyloxantin activity, as the study of Yehia et al. [12] demonstrated that celastrol efficiently STX biosynthesis in *S. aureus* through its effect on CrtM efficiently, confirmed by liquid chromatography-mass spectrometry (LC-MS) and molecular docking. In addition to its anti-pigment capability, celastrol exhibits significant anti-biofilm activity with its inhibitory effect on bacterial cell exopolysaccharides. Similarly, inhibition of STX upon celastrol treatment rendered *S. aureus* more susceptible to membrane targeting antibiotics. As a novel anti-virulent agent against *S. aureus*, Celastrol provides a prospective therapeutic role as a anti-pathogenic agent with multi-targets (**Table 1**).

3. Anti-biofilm activity of synthetic molecules

3.1 Anti-biofilm activity of meta-bromo-thiolactone (mBTL) via QS inhibition

In the study of O'Loughlin et al. [13], synthetic molecules were analyzed to prove their inhibitor effects on the two *P. aeruginosa* quorum-sensing receptors, LasR and RhlR. It was found that the most effective compound, is the meta-bromo-thiolactone (mBTL). It was also confirmed that both LasR and RhlR are partially inhibited by mBTL *in vivo* and *in vitro*; however, RhlR, not LasR, is the relevant *in vivo* target. Therefore, this work, that explores interference with QS, demonstrates that mBTL, an analogue of the native self-inducers of *P. aeruginosa*, suppresses the expression of genes encoding the virulence factor pyocyanin, on the one hand, and prevents biofilm formation on the other hand, which protects *C. elegans* and human lung epithelial cells from attack by *P. aeruginosa*. Taken together, these data about mBTL provide a strong argument for the efficacy of QS modulators for attenuation of QS-controlled phenotypes in pathogenic bacteria, such as biofilm formation (**Table 1**).

3.2 Anti-biofilm effect of malondialdehyde (MDA) via cell membrane injury

Malondialdehyde (MDA), one of the most representative reactive carbonyl species (RCSs) produced by lipid oxidation in bacteria [19] and in food [14], has received extensive attention recently. However, the inhibitory effect of MDA on microorganisms has received little attention. The study of Zhang et al. [10] proved the antibacterial effects of MDA on *S. xylosus* and *Lactiplantibacillus plantarum* with the MICs of 90 and 180 μg/ml, respectively. In addition, the antibacterial mechanisms of MDA on these two bacteria were associated with LDH activity changes as the LDH release is indicator of cell wall injury, accompanied with Ca^{2+} and Mg^{2+} leakage. Overall, the emission of Ca^{2+} and Mg^{2+} demonstrated that MDA enhanced the permeability of *S. xylosus* and *L. plantarum* cell membrane and further affected bacterial metabolism. In addition, MDA treatment induces cell membrane depolarization, indicating severe membrane damage with important impact on cell development and differentiation. This result has been confirmed by combination of CLSM and FEGSEM observations which have affirmed that MDA disrupts the cell membrane of *S. xylosus* and *L. plantarum*. It was also shown that MDA treatment significantly reduced the ATP concentration in *S. xylosus* and *L. plantarum*, suggesting that MDA may inhibit their growth by affecting the metabolic functions or cell membrane permeability of bacteria. Moreover, FT-IR studies showed that MDA might affect the molecular composition of *S. xylosus* and *L. plantarum* cells. These changes indicated the negative influence of MDA on cell membrane and cellular homeostasis [14].

4. Anti-biofilm Furanone activity through anti-QS activity

4.1 Furanones in general

Furanones are a family of structurally related molecules characterized by the presence of a five-membered heterocyclic furan ring. Furanones are available in a number of natural sources like marine and terrestrial plants, strawberries, coffee and fungi, and can also be chemically synthesized. Both natural and synthetic furanones have been shown to effectively inhibit QS, but synthetic furanones offer the possibility of precise control over compound structure and, therefore, control of any potential off-target effects [20, 21].

4.2 Anti-QS activity of natural furanones

Among the widely studied natural furanones, (5Z)-4-bromo-5-(bromomethylene)-3-butyl-2(5H)-furanone and Furanone 4-hydroxy-2,5-dimethyl-3 (2H)-furanone (HDMF), Furanone F202 show strong anti-biofilm activity by up to 55%. Based on the study of Ren et al. [22], a natural furanone, known as (5Z)-4-bromo-5- (bromomethylene)-3-butyl-2(5H)-furanone was demonstrated to attenuate biofilm formation in *E. coli*, reducing average biofilm thickness by 55%. Moreover, lower furanone concentrations (64.5 μM) significantly decreased *E. coli* swarming motility. On the other hand, investigations by Witsø et al. [23] into the impact of synthetic brominated furanones demonstrated that these compounds could also decrease *E. coli* biofilm wall thickness and surface area coverage by up to 50%. Brominated furanones, added at 50 μM, could suppress swarming motility and lower biofilm production by up to 40% in the foodborne pathogen *E. coli* 0103:H2. These important works [23] clearly demonstrated that natural furanones can interfere with QS processes and that the phenomenon could be used to combat virulence in human pathogens. Moreover, the study of Choi et al. [24] proved that natural furanones greatly reduce the production of *P. aeruginosa* virulence factors, including protease (up to 43%), chitinase and pyoverdine by almost 100% (**Table 2**) [24].

Furanones	Origin	Dose	MDR	Biofilm inhibition	Target	References
(5Z)-4-bromo-5-(bromomethylene)-3-butyl-2(5H)-furanone	Algae *Delisea pulchra*	164 μM 64.5 μM	*E. coli*	55% biofilm thickness reduction Swarming inhibition	QS process	[22, 25]
		[16.13–32.26 μM]	*Vibrio harveyi*	NM		
		64.5–322.5 μM	*P. aeruginosa* PAO1 and JB2	NM	i. increase in siderophore production ii. Protease reduction (43%) Chitinase, pyoverdine reduction (100%)	

Furanones	Origin	Dose	MDR	Biofilm inhibition	Target	References
Furanone 4-hydroxy-2,5-dimethyl-3(2H)-furanone (HDMF)	Variety of fruits	0.1 or 1 μM	*P. aeruginosa PAO1*	27.8% (0.1 μM) and 42.6% (1 μM) of biofilm inhibition	Reduced rhamnolipid (40.9%), pyocyanin (51.4%), LasA protease (53.8%) production	[24]
Furanone F202	Algae *Delisea pulchra*	50 μM	*E. coli* 0103:H2	50% biofilm inhibition	QS process	[23]

Table 2.
List of natural furanone inhibitors.

4.3 Anti-QS activity of synthetic furanones

The process of developing synthetic furanones began in the 1980s and it usually starts with relative simple compounds, such as dimethyl ketones and acetals or other straight forward organic precursors. Then, it is also possible to modify existing furanone compounds and add existing furanone structures, as it is highlighted in **Table 3**. Recently, the majority of research on furanones-mediated QS inhibitors has been conducted on the effects of these compounds on human pathogens, especially on the

Synthetic Furanones	Dose	MDR	Biofilm Inhibition	References
Tribromofuranone	50 μM	*S. enterica*	72% biofilm reduction Via Quorum sensing	[1]
		S. aureus	71% biofilm inhibition	
		C. albicans	51% biofilm inhibition	
Bis-4-methoxyphenylacetylene-substituted	NM	*E.coli* ATCC9637	31% biofilm inhibition	
Control monobromofuranone	NM	*P. aeruginosa PAO1*	75% biofilm inhibition	
Methyl-containing dibromofuranones	NM	*P. aeruginosa* PAO1 PAR7244	70% 44% biofilm inhibition	
Dibromofuranone Chloroiodofuranone	50 μM	i. *C. albicans M2396* ii. Mixed biofilm: *C. albicans M2396* with *P. aeruginosa PAO1*	92% biofilm inhibition	
Brominated Furanone C-30	3.125–50 μM.	*P. aeruginosa*	i. Near total prevention of pyoverdine production	[26] [21]

Synthetic Furanones	Dose	MDR	Biofilm Inhibition	References
			ii. Significant inhibition of iii. LasR iv. RhIR	
Furanone C-56	5 μg ml − 1 (28.5 μM)	*P. aeruginosa*	37% reduction in biofilm thickness	[26]

Table 3.
List of synthetic furanone inhibitors.

model organisms *E. coli* and *P. aeruginosa* [20]. It was the synthesis of a range of structurally diverse bromine-, chlorine and iodine-containing furanones using a variety of palladium-catalysed coupling reactions was recently described [1]. The finding of this study [1] is interesting, as furanone is an ideal QS disruptor. Various compounds from the furanone library were screened for their inhibitory effects on biofilm production in opportunistic human pathogens and were found to potently suppress bacterial biofilm formation in *S. enterica*, *S. aureus*, *P. aeruginosa*, and, to a lesser extent, *E. coli*. Compounds which inhibited biofilm formation do not generally impact bacterial growth, highlighting their potential as QS inhibitors. According to the Furanone Library [1], tribromofuranone was found to be the most active compound decreasing biofilm formation in *S. enterica* by 72% and *S. aureus* by 71% at a concentration of 50 μM, whereas methyl-substituted dibromofuranone was the most potent inhibitor, which reduces biofilm growth *P. aeruginosa* PAR7244 by 44% compared to a 70% reduction in PAO1. For *E. coli* biofilm, bis-4-methoxyphenylacetylene was the most active compound, which inhibits *E. coli* ATCC9637 biofilm growth by 31%. Moreover, it was tested whether synthesized furanones, with relevant anti-biofilm activity, were able to disturb mixed fungal-bacterial biofilms. It was confirmed that the chosen bromofuranones and chloroiodofuranones were initially subjected to testing for their effect on monospecific biofilms of *P. aeruginosa* and *C. albicans* with confocal laser scanning microscopy (CLSM). Thus, it was found that all of them decrease the biomass of both microorganisms within the mixed biofilms [1].

5. Anti-biofilm activity of natural compounds

Natural products exhibit higher structural and biochemical variety than synthetic compounds, making them useful for the advancement of anti-biofilm agents [27]. More recently, there have been reviews of bacterial products that include small molecules, enzymes, exopolysaccharides and isolated peptides displaying anti-biofilm activities toward different pathogens [28]. Moreover, several studies have demonstrated solid evidences that plants [29] and marine-derived products [15] are an excellent source to provide abundant natural compounds for the development of preventative and therapeutic agents against biofilm-based infections (**Table 1**).

5.1 Anti-biofilm activity of marine and bacterial-derived products

Concerning antibiofilm activity of marine-derived products, the study by Oluwabusola et al. [15] proved that psammaplin A and bisaprasin, isolated from marine sponges, could be a potent QS inhibitory agents by preventing *P. aeruginosa* PAO1 biofilm formation. The present results indicated that psammaplin a showed moderate to significant inhibition against QS gene promoters, with IC50 values ranging from 30.69 to 2.64 μM. In contrast, bisaprasin showed significant inhibition for both biosensor strains, with equal IC50 values. Hence, using marine sources to find novel QS inhibitors as antipathogenic drugs to combat antimicrobial resistance has high potentials. Concerning antibiofilm activity of bacterial-derived products, the study of Zhang et al. (2021) described a novel and effective anti-biofilm compound named maipomycin A (MaiA), which was isolated from the metabolites of a rare actinomycete strain *Kibdelosporangium phytohabitans* XY-R10. This compound demonstrated a broad spectrum of anti-biofilm activities against Gram-negative bacteria [8].

5.2 Anti-biofilm activity of thymol

The study of Valliammai et al. [17] demonstrated the anti-biofilm potential of thymol against methicillin resistant *S. aureus* (MRSA) by inhibition of staphyloxanthin biosynthesis. The staphyloxanthin inhibitory potential of thymol was assessed against MRSA in terms of quality and quantity. It was demonstrated that 100 μg/mL concentration of thymol brings about 90% of staphyloxanthin inhibition. In addition, it was confirmed that thymol treatment makes MRSA more susceptible to reactive oxygen species. Experimental analyses were also confirmed by molecular docking analysis and in vitro measurement of metabolic intermediates of staphyloxanthin. It was also revealed that thymol could possibly interact with CrtM, which is involved in staphyloxanthin biosynthesis to inhibit production. In addition, reduction in staphyloxanthin by thymol treatment increases the membrane fluidity and makes MRSA cells more susceptible to Polymyxin B, an antibiotic targeting membrane. Thus, the present study suggests thymol as a potential alternative to antibiotics to combat MRSA infections. It can also be used as adjuvant in antimicrobial treatments [17]. Likewise, Ndezo et al. [16] showed the synergistic effect of the anti-biofilm potential of thymol and piperine either alone or combined with three aminoglycoside antibiotics were evaluated against the biofilm of *K. pneumoniae*. Their effect were also tested on either formed or pre-formed biofilms. It was found that the minimal biofilm inhibition concentration (MBIC) of streptomycin was reduced 16- to 64-fold when associated with thymol, whereas the MBIC of kanamycin was decreased 4-fold when associated with piperine. In addition, the minimal biofilm eradication concentration (MBEC) values of streptomycin, amikacin, and kanamycin were 16- to 128-fold, 4- to 128-fold, and 8- to 256-fold higher than the planktonic minimum inhibitory concentration (MIC), respectively. Therefore, thymol, in combination with antibiotics, has shown a broad synergistic activity in both inhibiting biofilm formation and destroying pre-formed biofilm of *K. pneumoniae* [16].

The synergistic effects associated with the combination of thymol or piperine along with the three considered aminoglycoside antibiotics indicate that thymol and piperine are very promising agents for the development of new antibacterial combination therapies to combat biofilm-associated infections. The study by Valliammai et al. (2020) aimed to decrypt the molecular mechanism for the anti-biofilm activity of thymol toward MRSA and to evaluate the ability of thymol to enhance the antibacterial activity of rifampicin. Thymol markedly inhibited 88% of MRSA biofilm formation at 100 μg/mL and decreased MRSA adhesion to human plasma-coated glass, stainless steel, and titanium surfaces, as demonstrated by microscopic analysis. In fact, thymol reinforced the antibacterial efficacy and biofilm eradication of rifampicin against MRSA and also minimized the formation of persisters. Thus, the present study suggests that thymol is a very promising combinatory agent candidate to enhance the antibacterial activity of rifampicin for persistent MRSA infections [18].

6. Conclusion

Bacterial biofilms appear in many infections that are related to diverse medical implants and well defined body sites such as the urinary tract, lungs, wounds and their resistance to antimicrobial treatments is a serious problem in clinical settings. It is, therefore, imperative to study efficient solutions to this problem and to find an alternative to our current armory of antibiotics. The challenges related to biofilm infections have prompted researchers to seek a better understanding of the molecular mechanisms involved in biofilm formation, which has led to the identification of several steps in biofilm formation that could be targeted to eradicate these serious infections. Within this context, the combination of current antibiotics with potential anti-biofilm and anti-toxic agents that interfere with the QS without stimulating the incidence of resistance is a new therapeutic strategy aiming to reduce the antibiotic dosages. In this study, a screening of the most studied molecules with anti-biofilm activity, associated with or not with antibiotics, is performed. The different anti-biofilm molecules investigated here have various modes of action including (i) inhibition via interference in QS pathways by 3-PPA, AS10, mBTL, natural and synthetic furanones and natural compounds, (ii) adhesion mechanism, (iii) disruption of extracellular DNA, proteins, lipopolysaccharides, exopolysaccharides and secondary messengers involved in various signaling pathways like small molecule DGC-inhibitors of c-di-GMP signaling. As QS and c-di-GMP signaling govern the

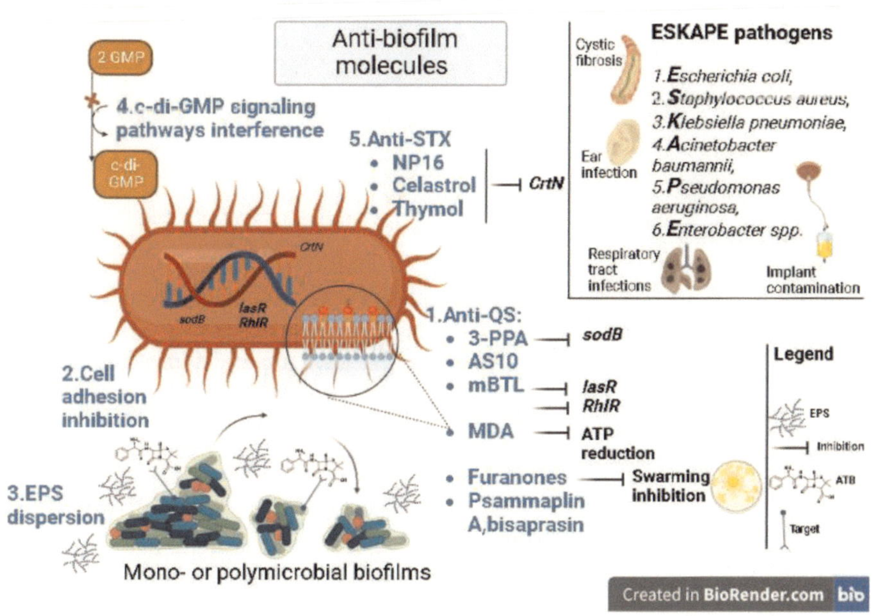

Figure 1.
Various modes of action of the studied anti-biofilm molecules in ESKAPE pathogenes.

production of virulence factors and some of the protective mechanisms operating in the biofilm mode, development of chemical compounds capable of preventing formation of biofilms by targeting these two major systems could be used to treat biofilm-associated infections. However, studies on the structural modifications on these molecules and their minimal effective concentration without posing harmful side effects should be made in future studies in order to improve their efficacy (**Figure 1**).

Abbreviations

STX	staphyloxantin
QS	quorum sensing
EPS	exopolysaccharides
MDA	malondialdehyde
ATB	antibiotics
3-PPA	3-phenylpropan-1-amine
c-di-GMP	cyclic dimeric guanosine monophosphate
mBTL	meta-bromo-thiolactone

Author details

Salma Kloula Ben Ghorbal*, Sana Dhaouadi, Sana Bouzenbila, Ameur Cherif and Ramzi Boubaker Elandoulsi
ISBST BVBGR-LR11ES31, Biotechnopole Sidi Thabet, University of Manouba, Ariana, Tunisia

*Address all correspondence to: salmakloula@yahoo.fr

References

[1] Gómez A-C, Lyons T, Mamat U, Yero D, Bravo M, Daura X, et al. Synthesis and evaluation of novel furanones as biofilm inhibitors in opportunistic human pathogens. European Journal of Medicinal Chemistry. 2022;**242**:114678

[2] Broncano-Lavado A, Santamaría-Corral G, Esteban J, García-Quintanilla M. Advances in bacteriophage therapy against relevant multidrug-resistant pathogens. Antibiotics. 2021; **10**(6):672

[3] Madalina, Mihai M, Maria Holban A, Giurcaneanu C, Gabriela Popa L, Mihaela Oanea R, Lazar V, et al. Microbial biofilms: Impact on the pathogenesis of periodontitis, cystic fibrosis, chronic wounds and medical device-related infections. Current Topics in Medicinal Chemistry. 2015;**15**(16):1552-1576

[4] Roy R, Tiwari M, Donelli G, Tiwari V. Strategies for combating bacterial biofilms: A focus on anti-biofilm agents and their mechanisms of action. Virulence. 2018;**9**(1):522-554

[5] Zhao X, Yu Z, Ding T. Quorum-sensing regulation of antimicrobial resistance in bacteria. Microorganisms. 2020;**8**(3):425

[6] Yin L, Zhang P-P, Wang W, Tang S, Deng S-M, Jia A-Q. 3-Phenylpropan-1-amine enhanced susceptibility of Serratia marcescens to Ofloxacin by occluding quorum sensing. Microbiology Spectrum. 2022;**10**(5):e01829-e01822

[7] De Brucker K, Delattin N, Robijns S, Steenackers H, Verstraeten N, Landuyt B, et al. Derivatives of the mouse cathelicidin-related antimicrobial peptide (CRAMP) inhibit fungal and bacterial biofilm formation. Antimicrobial Agents and Chemotherapy. 2014;**58**(9): 5395-5404

[8] Zhang J, Liang X, Zhang S, Song Z, Wang C, Xu YJ. Maipomycin a, a novel natural compound with promising anti-biofilm activity against gram-negative pathogenic bacteria. Frontiers in Microbiology. 2021;**11**:598024

[9] Zhang Y, He X, Cheng P, Li X, Wang S, Xiong J, et al. Effects of a novel anti-biofilm peptide CRAMP combined with antibiotics on the formation of Pseudomonas aeruginosa biofilms. Microbial Pathogenesis. 2021;**152**:104660

[10] Zhang Y, Cheng P, Wang S, Li X, Peng L, Fang R, et al. Pseudomonas aeruginosa biofilm dispersion by the mouse antimicrobial peptide CRAMP. Veterinary Research. 2022;**53**(1):80

[11] Gao P, Davies J, Kao RYT. Dehydrosqualene desaturase as a novel target for anti-virulence therapy against Staphylococcus aureus. MBio. 2017;**8**(5): e01224-e01217

[12] FA-ZA Y, Yousef N, Askoura M. Celastrol mitigates staphyloxanthin biosynthesis and biofilm formation in Staphylococcus aureus via targeting key regulators of virulence; in vitro and in vivo approach. BMC Microbiology. 2022;**22**(1):1-18

[13] O'Loughlin CT, Miller LC, Siryaporn A, Drescher K, Semmelhack MF, Bassler BL. A quorum-sensing inhibitor blocks Pseudomonas aeruginosa virulence and biofilm formation. Proceedings of the National Academy of Sciences. 2013;**110**(44): 17981-17986

[14] Zhang Q, Jia S, Ding Y, Li D, Ding Y, Zhou X. Antibacterial activity and

mechanism of malondialdehyde against staphylococcus xylosus and Lactiplantibacillus plantarum isolated from a traditional Chinese dry-cured fish. Frontiers in Microbiology. 2022;**13**: 979388

[15] Oluwabusola ET, Katermeran NP, Tan LT, Diyaolu O, Tabudravu J, Ebel R, et al. Inhibition of quorum sensing system in Pseudomonas aeruginosa by Psammaplin a and Bisaprasin isolated from the marine sponge Aplysinella rhax. bioRxiv. 2021:2021.01. 22.427779

[16] Bisso, Ndezo B, Tokam Kuaté CR, Dzoyem JP. Microbiology M. synergistic antibiofilm efficacy of thymol and piperine in combination with three aminoglycoside antibiotics against Klebsiella pneumoniae biofilms. Canadian Journal of Infectious Diseases and Medical Microbiology. 2021;**2021**: 1-8

[17] Valliammai A, Selvaraj A, Muthuramalingam P, Priya A, Ramesh M, Pandian SK, et al. Staphyloxanthin inhibitory potential of thymol impairs antioxidant fitness, enhances neutrophil mediated killing and alters membrane fluidity of methicillin resistant Staphylococcus aureus. Biomedicine & Pharmacotherapy. 2021;**141**:111933

[18] Valliammai A, Selvaraj A, Yuvashree U, Aravindraja C, Pandian K. sarA-dependent antibiofilm activity of thymol enhances the antibacterial efficacy of rifampicin against Staphylococcus aureus. Frontiers in Microbiology. 2020;**11**:1744

[19] Salma KBG, Lobna M, Sana K, Kalthoum C, Imene O, Abdelwaheb C. Antioxidant enzymes expression in Pseudomonas aeruginosa exposed to UV-C radiation. Journal of Basic Microbiology. 2016;**56**(7): 736-740

[20] Proctor CR, McCarron PA, Ternan NG. Furanone quorum-sensing inhibitors with potential as novel therapeutics against Pseudomonas aeruginosa. Journal of Medical Microbiology. 2020;**69**(2):195-206

[21] Markus V, Golberg K, Teralı K, Ozer N, Kramarsky-Winter E, Marks RS, et al. Assessing the molecular targets and mode of action of furanone C-30 on Pseudomonas aeruginosa quorum sensing. Molecules. 2021;**26**(6):1620

[22] Ren D, Sims JJ, Wood TK. Inhibition of biofilm formation and swarming of Escherichia coli by (5Z)-4-bromo-5-(bromomethylene)-3-butyl-2 (5H)-furanone. Environmental Microbiology. 2001;**3**(11):731-736

[23] Witsø IL, Benneche T, Vestby LK, Nesse LL, Lönn-Stensrud J, Scheie AA, et al. Thiophenone and furanone in control of Escherichia coli O103: H2 virulence. Pathogens and Disease. 2014;**70**(3):297-306

[24] Choi S-C, Zhang C, Moon S, Oh Y-S. Inhibitory effects of 4-hydroxy-2, 5-dimethyl-3 (2H)-furanone (HDMF) on acyl-homoserine lactone-mediated virulence factor production and biofilm formation in Pseudomonas aeruginosa PAO1. Journal of Microbiology. 2014;**52**: 734-742

[25] Ren D, Zuo R, Wood TK. Quorum-sensing antagonist (5Z)-4-bromo-5-(bromomethylene)-3-butyl-2 (5H)-furanone influences siderophore biosynthesis in pseudomonas putida and Pseudomonas aeruginosa. Applied Microbiology and Biotechnology. 2005; **66**:689-695

[26] Hentzer M, Wu H, Andersen JB, Riedel K, Rasmussen TB, Bagge N, et al. Attenuation of Pseudomonas aeruginosa virulence by quorum sensing inhibitors. The EMBO Journal. 2003;**22**(15): 3803-3815

[27] Genilloud O. Actinomycetes: Still a source of novel antibiotics. Natural Product Reports. 2017;**34**(10):1203-1232

[28] Khan F, Oloketuyi SF, Kim Y-M. Diversity of bacteria and bacterial products as antibiofilm and antiquorum sensing drugs against pathogenic bacteria. Current Drug Targets. 2019;**20**(11):1156-1179

[29] Song X, Xia Y-X, He Z-D, Zhang H. A review of natural products with anti-biofilm activity. Current Organic Chemistry. 2018;**22**(8):789-817

Chapter 9

Important Advances in Antibacterial Nanoparticle-Mediated Photodynamic Therapy

Sandile Phinda Songca

Abstract

Earlier applications of photodynamic therapy (PDT) were accomplished by direct or intravenous injection of the photosensitizer, followed by preferential accumulation in cancerous tissues after systemic circulation. Nowadays, nanoparticles are used as carriers and delivery systems, which also facilitate combinations of PDT with other non-invasive technologies. PDT has expanded to disease types other than cancers. Nanoparticle-mediated target specific PDT can reduce the emergence of resistance, and has introduced chemotherapy combinations with PDT, and potential repurposing of chemotherapy drugs that are being used less because of resistance. The novel discoveries of inorganic and organic dye nanoconjugate photosensitizers discussed in this chapter have enhancement PDT efficacy. This review describes the type I and II mechanisms of PDT, some of the first- and second-generation photosensitizers in the market, and the roles played by nanomaterials across the PDT clinical translation value chain. It discusses nanoparticles as delivery systems for photosensitizers, smart stimulus-responsive, and disease-targeting nanoparticles, focusing on folate, glycan-based, pH, and external stimulus-responsive targeting. Well-known in anticancer applications, folate targeting is now debuting in antibacterial applications. Other targeting technologies are discussed. Nanoparticles applications as agents for combining PDT with other therapies are discussed. The World Health Organization has identified PDT as a promising new technology.

Keywords: antibacterial photodynamic therapy, photodynamic antimicrobial chemotherapy, nanoparticle-mediated photodynamic therapy, bacterial resistance, bacterial cell specificity, selectivity, drug carrier, drug delivery

1. Introduction

From a clinical perspective, photodynamic therapy (PDT) may be defined as a treatment that involves the application of light energy in a disease-affected area where there is a sufficient concentration of the photosensitizer (PS). PDT destroys disease cells only upon activation by light, provided there is a sufficient oxygen concentration in the disease. PSs are generally activated using laser light of a wavelength that is absorbed by the PS. They are nontoxic compounds that become toxic upon light activation. Clinical PDT is widely used against psoriasis; cancers of the skin, lung, brain, bladder, pancreas, bile duct, esophagus, and head and neck; as well as other diseases such as acne and age-related macular degeneration. Additionally, antimicrobial photodynamic therapy (aPDT) is used to treat bacterial, fungal, and viral infections. It has been established from several studies that there is an immune response to PDT that can further enhance its efficacy. From a mechanistic point of view, PDT involves the excitation of the PS to its singlet excited state upon absorption of light of a frequency that matches the absorption spectrum of the PS, followed by intersystem crossing to the triplet state, which is the state in which the PS either transfers energy to normal triplet state oxygen to produce excited singlet state oxygen or reacts with biomolecules, causing damage to cells. Singlet oxygen production is the most effective PDT pathway. It takes place under conditions of oxygenation and is referred to as type II mechanism. In contrast, the direct reaction of the excited PS with biomolecules, referred to as type I mechanism, predominates under conditions of hypoxia, because there is not sufficient oxygen for type II mechanism.

In the aPDT approach, absorbed light energy is used to achieve the bactericidal or bacteriostatic effect through these two critical molecular PS-mediated mechanisms. Here the type I mechanism involves much radical formation through hydrogen transfer from the PS directly to biomolecules, and the type II mechanism proceeds via oxygen photosensitization to produce a series of oxygen-based molecular species known as reactive oxygen species (ROS), which includes singlet oxygen, oxygen radicals, and hydroxide radicals and radical anions. All ROS react with biomolecules, causing irreversible damage [1]. The Jablonski diagram shown in **Figure 1** illustrates the two mechanisms. The irreversible chemical reactions that alter the functionality of biomolecules in bacterial cells and the extracellular polymeric substance (EPS) matrix [3], regardless of whether these biomolecules are cellular, EPS matrix components, or some other functional constituents of the biofilm [4], have been extensively studied. Many of these studies conclude that aPDT increases intracellular ROS and reduces the strength of the EPS matrix and the metabolic activity of the pathogen cells in the matrix [5].

Nanoparticles may be defined as ultra-small particulate materials with one of the dimensions of the particles up to 100 nanometers. Metallic nanoparticles like metal chalcogenide and silica nanocomposites have been reported. Self-assembled phospho-lipid porphysome vesicles [6] and phthalocyanine-based porphysome-like nanostructures [7] are very common PSs for PDT. Organo-inorganic nanomaterials comprise organic and inorganic nucleated heterocyclic aromatic organic compounds in self-assembled nanoparticle (NP) formations. Recently, metal organic frameworks (MOFs) have emerged, in which the linking organic molecules are PS molecules such as phthalocyanines and porphyrins [8, 9]. Typical applications of these PS molecules in PDT include antiviral, antibacterial, antifungal, anticancer, pest control, and environmental sanitization [2]. Several combination therapies, gene therapies, immunotherapies, and checkpoint blockade immunotherapies, in which these molecules are used as integral parts of PS nanoconjugate systems, have been widely reported. These nanoconjugates are widely reported in pharmaceutical formulation; controlled, stimulus-responsive, and slow drug release; enhancement of bioavailability; combination therapies; and enhancement of therapeutic efficacy, using a range of techniques such as nano-crystallization and self-assembly. Nanomaterials are found at every node of the therapeutic value chain and drug development pipeline, from

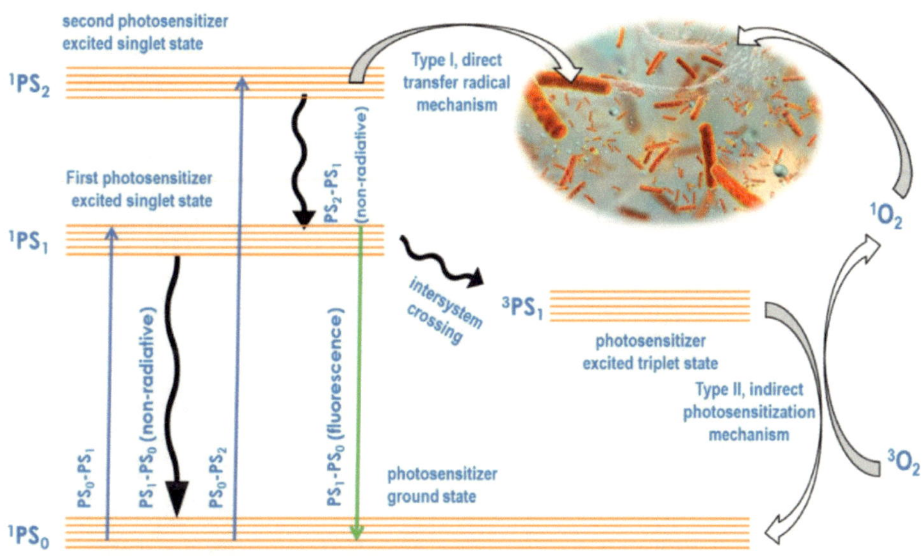

Figure 1.
Jablonski diagram to illustrate the aPDT type I and II mechanisms. Reproduced from Songca and Adjei [2] under the creative commons attribution license 4.0.

basic drug research and development, through 2D and 3D evaluations in vitro, finally arriving at the preclinical studies in vivo, pharmaceutical formulation, applications in clinical trials, and drug administration in clinical therapy.

The challenge of incorporating PS molecules that are used for PDT and other drug molecules that are used as antibacterial chemotherapeutic agents, into innovative nanoconjugate systems, in designing them to act as carriers and delivery vehicles of the PS and drug molecules, and act as systems that respond to internal or external stimuli, once they are internalized into disease cells, is an important preoccupation of scientists in nanomaterial-mediated PDT. The purpose of incorporating PS molecules and antibacterial chemotherapeutic drug molecules into innovative nanoconjugate systems is to ensure their inertness and non-toxicity while in systemic circulation. The purpose of building internal and external stimuli responsiveness is to ensure that they are released only when the nanoconjugate is inside the target disease cell or site, when the stimuli of the internal environment of these cells or sites trigger their release, or when an external stimulus is applied (**Figure 2**).

Incorporating small molecules of antibacterial drugs as components of nanoconjugates presents many advantages in efficacy improvement. These include pharmacokinetic navigation of various physiological barriers and reduction of many of their side effects, including the development of bacterial resistance. Most of the self-assembly reactions used are conducted in aqueous media to form NPs composed of small, potent drug molecules. Most nanoconjugates are easy to fabricate; they can deliver high concentrations of their drug molecule cargo to the disease microenvironment and intracellular environment of the disease cells. Given the facile pharmacokinetic navigation of the systemic barriers by these drugs when capped or otherwise encapsulated in nanoconjugate form, they have great potential to reduce or eliminate their side effects because they are only released at the disease site and, in the most innovative designs, they are released only once they are inside the disease cells, and are not released inside normal host tissue cells.

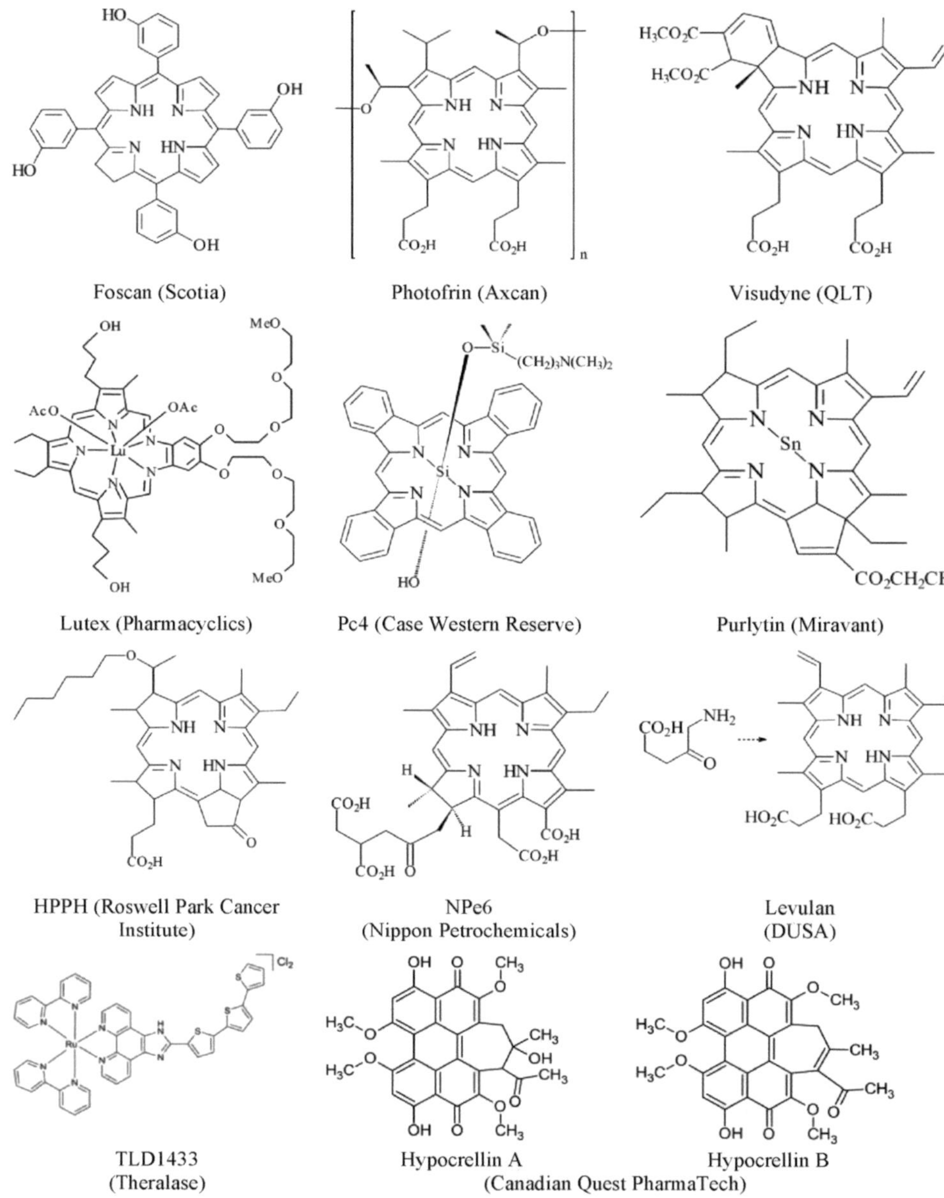

Figure 2.
Chemical structures of Foscan, Photofrin, Visudyne, Lutex, Pc4, Purlytin, HPPH, NPe6, Levulan, TLD1433, Hypocrellin a, and Hypocrellin B. The pharmaceutical companies are indicated in brackets.

Most PSs used in PDT are organic molecular chromophores that are capable of transferring electromagnetic radiation energy to oxygen to form ROS in situ [10]. However, the new inorganic NP PSs that have been discovered are showing good versatility because in addition to being used for transport and delivery of the PSs, they can also act as PDT PSs, photothermal therapy, and magnetothermal therapy agents in combination therapies with PDT. For example, an inorganic NP PS consisting of Fe_2O_3, and CuS, which also acts as a PS and therefore, possesses photothermal and magnetothermal conversion, in addition to PDT capabilities, as reported by Curcio et al. [11]. The nanoconjugate demonstrated its capabilities in a tri-therapeutic combination involving photothermal hyperthermia therapy (PTT), PDT and magnetic hyperthermia therapy (MH), in which the iron oxide shell is responsible for MH, and the copper sulfide multi-core is responsible for PTT and PDT. In their review, Zhang et al. [12] identified carbon-based inorganic nanomaterials such as dots, fullerenes, nanotubes, graphene oxide semiconductor nanomaterials such as zirconium and titanium oxides, and defective nanomaterials such as oxides of ruthenium and zinc, as some of the inorganic NPs that generate ROS upon photo irradiation. Conjugation of these nanomaterials with the organic dye type of PS results in efficacious nanoconjugates in combination therapies. For example, the conjugation of copper sulfide NPs with chlorin-e6 produced a potent PDT-and-PTT combination agent because both the core and shell materials produce ROS [13].

Examples of organic PSs that have clinical approval include Foscan from Scotia [14], Visudyne from QLT [15], Lutex from Pharmacyclics [16], Pc4 from Case Western Reserve [17], Purlytin from Miravant [18], NPe6 from Nippon [19], HPPH from Roswell Park Cancer Institute [20, 21], Amino Laevulinic Acid from DUSA [22], Hypocrellin Photosensitizer SL052 from Canadian Quest PharmaTech [23], and TLD1433 from Theralase [24]. Examples of inorganic PSs include sulfides of molybdenum, zinc, copper, iron, silver, and bismuth [25]. Nanostructured MOFs [26] and metal complexes with organic ligands [23], on the other hand, may therefore be considered among the wide and increasing variety of organic–inorganic hybrid nanostructured PDT PSs [27–29].

2. Purpose statement

This paper presents the roles played by nanomaterials across the therapeutic value chain, from basic research to clinical applications, using examples from therapeutic technologies and their clinical applications against many bacterial and fungal diseases. The paper adopts an approach of considering therapeutic applications of nanotechnology in treating bacterial diseases and the nanomaterial-based therapeutic technologies applied in treating them, discussing the details of these applications and the technologies that define them. Using photosensitization type I and II mechanisms by which ROS are produced in the disease microenvironment, and the subsequent oxidative initiation of apoptosis and necrotic cell death, the purpose of this paper is pursued by discussing specific examples. The purpose of this approach is to provide the fundamental mechanistic basis of the technology and its combinations, an overview of its state-of-the-art from the current research and the historical viewpoint, the expansion of its scope, the enhancement of its efficacy and disease targeting, and the role of nanotechnology in these developments. This paper also aims to discuss potential areas of further research and innovation as indicated by gaps in the basic research and clinical translation literature.

3. Nanoparticles as carrier and delivery systems for photosensitizers

The use of NPs as carriers and delivery agents for PSs and other drugs has gained much attention [30, 31] and has demonstrated the enhancement of stability, solubility, administration, target delivery, specificity, selectivity, and toxicity reduction [32, 33]. Research on NPs as carriers of PSs has demonstrated that the enhancement of PDT is due to the enhancement of PS drug delivery and cellular uptake and retention [34, 35]. Due to the ultra-small size of NPs, they have large surface-to-volume ratios [36]. This allows them to absorb large quantities of the PS on their surface [37], which promotes the target tissue and cellular uptake [38] once they reach the target site and cells. In addition, this PS drug delivery mechanism can also enhance selectivity for the disease site and cells over host tissue sites or cells. The absorption of PSs on the surface of NPs increases NP stability while in systemic circulation [39]. This severely limits undesirable side effects of both the NP and PS, such as toxicity in the absence of light. PS-capped NPs generally have improved solubility in hydroxylic media, thus enhancing the administration of the nanoconjugate [40]. The foregoing discussion describes the encapsulation of NPs by PSs. It may be illustrated using the example of encapsulation of magnetic NPs with heparin-pheophorbide-A, as reported by Li et al. [41], shown in **Figure 3**, in which the aminopropyl triethoxysilane functionalized iron oxide NPs are encapsulated with heparin–pheophorbide-A by conjugation of the functionalized NPs. The encapsulation of NPs by PSs is one of the most effective and therefore most widely reported strategies for using NPs as carrier and delivery systems of PSs for use in PDT.

The PS may be covalently linked or adsorbed onto the surface of the NP. For example, a near-infrared absorbing and disulfide functionalized bacteriochlorophyll-a-based PS was covalently anchored onto the surface of gold NPs for anticancer PDT, using gold surface–sulfide dative covalent bonding of the disulphide functional group of the PS [42]. The researchers found that in comparison to the free bacteriochlorophyll-a-based PS, the gold-PS nanoconjugate remained in systemic circulation for longer and showed increased tumor accumulation, cancer cell uptake and retention. This nanoconjugate is illustrated in **Figure 4**.

In this case, the linker is a functional group of the PS. The covalent anchoring of Rose Bengal on the surface of silica NPs, however, was accomplished by functionalizing the NP surface with amino groups followed by covalent linking of the PS, via the formation of amide covalent bonds between the carboxylic acid functional group of Rose Bengal and the amino groups of the silica capping shell [43]. This is illustrated in **Figure 5**.

The nanoconjugate inactivated gram-positive Methicillin-resistant *S. aureus* and *Staphylococcus epidermidis*, indicating that this method of PS conjugation has great potential for aPDT applications.

The encapsulation of PSs in the core of organic NPs such as liposomes and micelles has emerged as a powerful way of enhancing PS delivery [44]. This is a

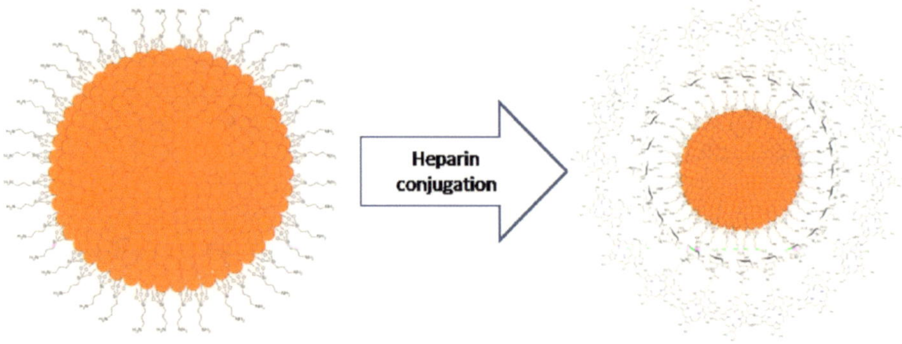

Figure 3.
Heparin–pheophorbide-a conjugation of aminopropyltriethoxysilane functionalized nanoparticles.

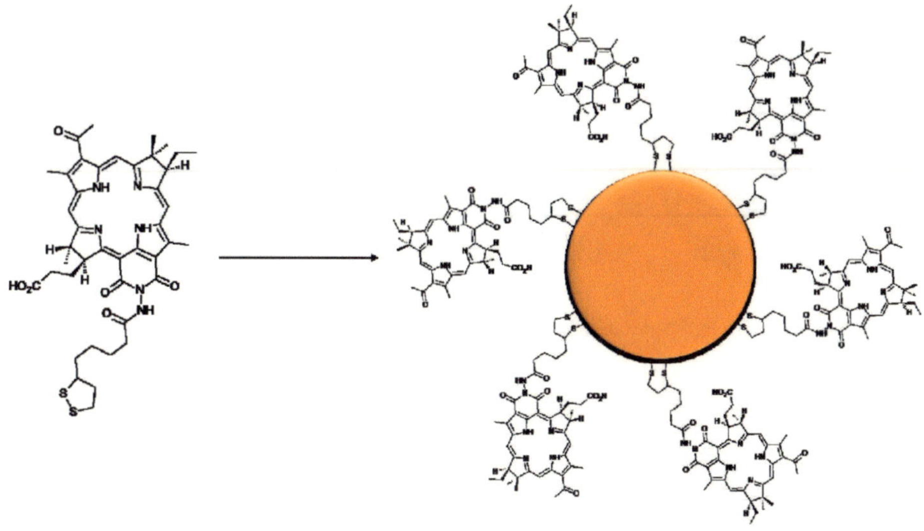

disulfide functionalized bacteriochlorophyll-a bacteriochlorophyll-a-gold nanoconjugate

Figure 4.
Covalent binding of bacteriochlorophyll-a-based PS onto the surface of gold nanoparticles.

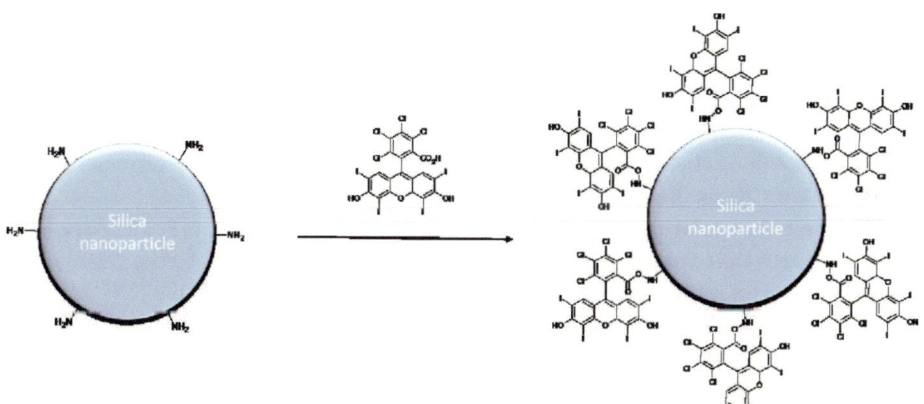

amino-functionalized silica nanoparticle amide covalently linked Rose Bengal

Figure 5.
Amide bond covalent binding via amino-functionalized silica nanoparticles.

versatile approach because, while hydrophilic PSs are encapsulated in water-in-oil organic NPs, hydrophilic PSs are encapsulated in oil-in-water organic NPs [45, 46]. This is due to the respective structures of the oil-in-water and water-in-oil NPs. The constituent molecules of these organic NPs, known as micelles, are phospholipids, which self-assemble with the alignment of their hydrophilic heads and hydrophobic tails. Water-in-oil organic NPs align with hydrophilic heads in the interior of the NP, thus encapsulating hydrophilic PSs in an aqueous medium, while oil-in-water NPs align with hydrophobic tails in the interior of the NP, thus encapsulating hydrophobic PSs in an organic medium [47]. Unlike micelles, which have a single layer of phospholipids, liposomes have a double layer of phospholipids with an aqueous

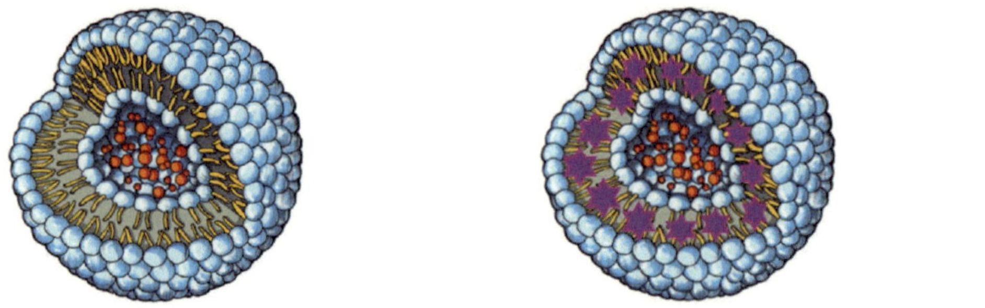

Figure 6.
Liposome with hydrophilic core hydrophobic bilayer, two partitions for hydrophilic and hydrophobic PSs respectively.

core, encapsulating hydrophilic PSs and a hydrophobic phospholipid bilayer that can accommodate large quantities of hydrophobic PSs [48]. The structures of liposomes are illustrated in **Figure 6**.

To overcome the lipophilicity of porphyrins that limits their water solubility, protoporphyrin IX was conjugated with oleylamine to enhance its solubility in the liposomal bilayer [49]. **Figure 6b** and 7 illustrate the liposomal bilayer incorporation of the hydrophobic PS. The in vitro anticancer studies of the liposome incorporated PS showed that it significantly reduced the viability of HeLa and AGS cancer cell lines. Bilayer incorporation of the PS was also observed with temoporphyrin [50].

In contrast, the water-soluble PSs like Methylene Blue, Neutral Red, and Rose Bengal are encapsulated into the inner aqueous core of the liposome [51]. The liposomal encapsulation of these PSs, which was evaluated by gel filtration chromatography using Sephadex 100, is illustrated in **Figure 6a**.

Unlike the encapsulation of NPs discussed above, mesoporous NPs (**Figure 8**) have micro-pores that are large enough to absorb large quantities of PSs, permeating through their entire nanostructures. Mesoporous silica [53] and MOFs [54] are among the most widely studied mesoporous NPs. The advantage of using mesoporous silica NPs as PS carriers and delivery systems is that they are biocompatible and safe to use [55]. Mesoporous silica NPs are fabricated to assume several 3D structures, which enable loading and the control of NP release at the target site, and surface functionalization, depending on the synthetic methodology [56]. Dendritic mesoporous silica nanostructures have now emerged with highly porous nanostructures and high loading capacity due to their large pore sizes [57]. MOFs can absorb large quantities of PSs in their large pore sizes that can be hydrophobic or hydrophilic depending on the organic molecule linkers and the linked metal cations. As a result, they can absorb hydrophilic PSs in hydrophilic pore sites and hydrophobic PSs in hydrophobic pore sites [58]. Although most MOF pores tend to be hydrophobic due to the hydrophobic organic molecule linkers, the design and self-assembly of hydrophilic MOFs have been reported for the absorption of hydrophilic molecules such as glycol peptides [59]. Although, in theory, such MOFs can also be used to absorb hydrophilic PSs, to the best of our literature search, such research has yet to be reported.

A new type of MOF consisting of porphyrins or phthalocyanines as the organic linkers has been reported to absorb large quantities of oxygen, thus alleviating hypoxia in PDT and acting as PSs by ROS generation [60]. Furthermore, to ameliorate the tissue penetration challenge of normal light energy used in PDT, porphyrin-based MOFs

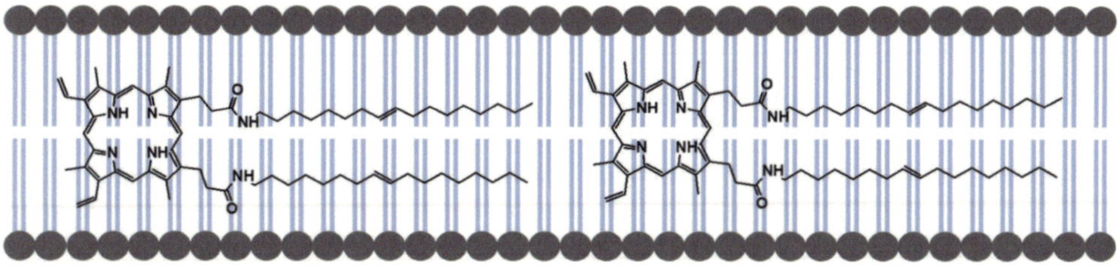

Figure 7.
Liposomal bilayer incorporation of the oleylamine conjugated protoporphyrin IX.

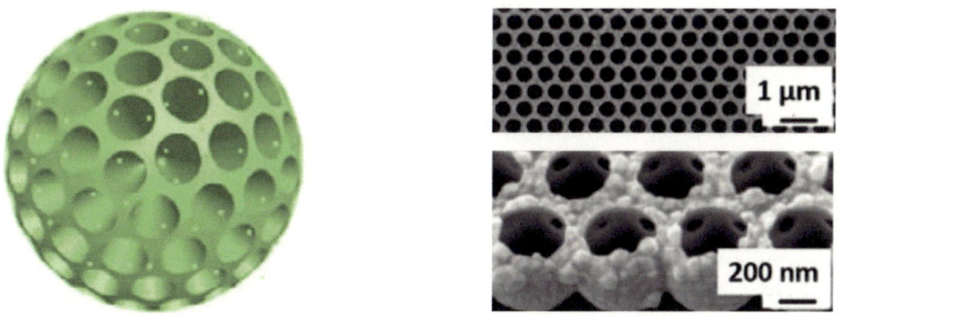

(a) illustration of mesoporous silica nanoparticle (a) magnification of metal-organic-frameworks pores

Figure 8.
Mesoporous silica nanoparticles and metal-organic-frameworks are highly porous nanomaterials. Copied from Zhang and Chang [52] under the creative common attribution license 4.0.

have been reported, which absorb X-rays and transfer the energy to the porphyrin linkers for oxygen sensitization to generate ROS [61]. Other mesoporous nanomaterials developed for use in PDT include mesoporous carbon and titanium NPs. For example, oxygenated perfluoro hexane was loaded onto the mesoporous carbon NP channels for antibacterial applications in combining PTT and PDT [62]. Mesoporous titanium oxide NPs have been developed for overcoming drug resistance in a combination therapeutic approach involving disease targeting and drug delivery in PDT [63].

4. Smart targeting nanoparticles in antimicrobial photodynamic therapy

Microbial infectious diseases, especially those due to bacterial and fungal infections, initially affect specific areas and may subsequently spread throughout the entire organism [64]. Therefore, in order to arrest bacterial pathogenesis, treatment modalities that identify and target affected areas, sites, and cells are preferred. The importance of using smart targeting NPs in aPDT emanates from the desire to direct such treatment to disease-affected areas, sites, and cells, with minimal or no negative effects on normal host tissue cells [65]. When the disease-infected site is external, and the aPDT treatment is topical rather than systemic, the purpose of smart NP-mediated targeting is to enhance selectivity for the infective microbial cells and their supporting EPS matrix over normal host tissue cells. In cases of deep tissue or systemic infection, however, such targeting has a general purpose of selectivity for disease cells [66]. Several methods have been reported to enhance selectivity for the disease cells over normal host tissue cells.

5. Aptamer-based targeting

Nanoconjugates functionalized with disease cell-specific aptamers have been reported to enhance specificity for microbial pathogens. Aptamers are single strands of intact sequences of nucleic or xeno nucleic acids. Because of their high affinity, selectivity, and specificity for specific microorganism targets, aptamers are selected and prepared, typically using the SELEX procedure [67] and used for NP functionalization [68]. The use of aptamers for the targeted delivery of anticancer drugs and PSs has been ubiquitously studied [69]. However, the use of aptamers for the targeted delivery of antibacterial drugs and PSs in aPDT has recently attracted attention [70]. Disease cell targeting aPDT applications may be illustrated with the studies of a DNA-aptamer-functionalized nanographene oxide as a targeted nanomaterial-mediated bio-theragnostic approach against *Porphyromonas gingivalis*, a pathogenic periodontitis constituent of the periopathogenic complex [71]. Following synthesis and characterization, the nanographene oxide was functionalized with an aptamer [72], which was selected using the SELEX procedure [73, 74]. Using fluorescence flow cytometry, this study showed that graphene oxide NPs, functionalized with the DNA aptamer, enhanced target specificity of the nanoconjugate for *P. gingivalis* disease cells. In a similar study, aptamer-functionalized emodin NPs showed binding specificity and enhanced antibacterial activity against *Enterococcus faecalis* [75]. Regarding the applications of aptamers in aPDT, literature reviews have indicated that the effect of aptamers goes beyond disease cell targeting to include bactericidal and biofilm disruptive effects [76], suggesting that in addition to targeting specific bacterial pathogens, the aptamer-functionalized nanoconjugates could also exhibit bactericidal and biofilm disruptive effects.

6. Glycan-based targeting

Evidence that the carbohydrate-based polysaccharide polymers found on bacterial cells, also known as glycans, can form the basis for bacterial targeting has been presented [77]. There are glycan-recognizing and binding proteins on bacterial target host cell surfaces, known as lectins. These protein molecules are recognized by the glycan structures on bacterial target host cell surfaces where the bacteria attach for host cell invasion [78]. The antibacterial macrophage strategy involves the initial attachment to the bacterial cell surface, followed by the delivery of depolymerases and lysins to degrade the bacterial cell wall-based glycans [79]. Similarly, the bacterial glycan cell targeting technology is based on extensively lectin-functionalized nanoconjugate systems that attract and selectively bind to bacteria with high binding affinity, delivering their antibacterial cargo, such as antibiotic chemotherapy drugs and PDT PSs, yet maintaining host microenvironment biocompatibility [80, 81].

7. Smart stimulus-responsive nanoparticles in antimicrobial photodynamic therapy

Further enhancement of selectivity for the disease over host tissue cells can be achieved if the aPDT toxicity of the drug and PS molecules is controlled in such a way that they are only toxic on target and are benign elsewhere. As a result, substantial research has been dedicated to developing stimulus-responsive aPDT. Two approaches have emerged to achieve this. In the one approach, nanoconjugate systems have been cleverly designed and fabricated to respond to the pH and redox potential difference between normal host tissue cells and the extracellular environment, on the one hand, and the intracellular environment of bacterial disease cells, on the other hand. In bacterial cells and the extracellular bacterial microenvironment, the pH drops by nearly 2–3 compared to normal host tissue cells and the usual host tissue extracellular microenvironment [82]. Therefore, systems have been cleverly designed in which the drug and PS molecules are covalently bound by functional groups that are cleaved upon the pH drop as they enter the disease cells. Due to the pH differential, this stimulus responsiveness selects only disease cells to deliver their drug and PS cargo and withholds it anywhere else.

Utilizing the pH differential, the PS curcumin was incorporated into the zeolitic imidazolate framework-8, ZIF-8, which disassembles at low pH, releasing the PS. The zinc ions released from the MOF increased the porousness of the bacterial cell membrane, causing the enhanced production of ROS in the extracellular environment, which resulted in bacterial cell membrane disruption and damaged appearance of the bacteria under the electron microscope [83]. Therefore, the authors concluded that pH-sensitive MOF-mediated bacterial cell targeting might be a promising aPDT strategy. A similar study showed pH-responsive delivery of ammonium methylbenzene blue incorporated into the ZIF-8 [82]. Clearly, the MOF strategy is an important approach to pH-sensitive drugs and PS release in aPDT. The technology of encapsulation of the PS in organic NPs has also been studied in pH-responsive targeting. For example, Chlorin e6-encapsulated pH-sensitive charge-conversion polymeric NPs were used to target *E. coli* infection in low pH urinary tract environments, with more than two-fold efficacy enhancement [84]. Additionally, liposomal encapsulation of PSs can be tuned to be pH-sensitive by formulation of the composition of the phospholipids that form the liposomal bilayer. For example, encapsulation of Chlorin-e6 into pH-sensitive liposomes fabricated by varying the composition of dipalmitoyl phosphatidylcholine, cholesterol, and dimethyl dioctadecyl ammonium chloride in chloroform resulted in selective penetration into the cytoplasm of *E. coli* [85].

Nanoconjugate systems have also been designed that respond to externally applied physical stimuli, such as MH, PTT, and US. The high preference for MH and ultrasound (US) is due to their unlimited tissue penetration depth compared to the limited tissue penetration depth of light, even in the therapeutic window. Utilizing external stimuli may be illustrated with the combination of MH with PDT by encapsulating magnetic iron oxide NPs in the liposome aqueous core and organic PSs in the hydrophobic liposomal bilayer (**Figure 9a**). The PS is released upon applying a high frequency alternating magnetic field (**Figure 9b**), which elevates the temperature to 42–45°C, disassembling the liposome and releasing the PS (**Figure 9c**) from the liposomal bilayer [86]. Encapsulating plasmonic and photo-responsive NPs also achieves the release of the PS in the same way, upon the application of light to elevate the temperature by the photothermal mechanism. Used to target cancer cells in experimental studies, this approach eradicated all cancer cells in an *in vitro* study and completely ablated the solid tumors *in vivo* [87].

Interestingly, to the best of our literature search, studies of MH in combination with PDT have not been reported, although studies on the antibacterial effects of static magnetic fields have been conducted. For example, applying an external magnetic field caused magnetic NPs to move deep into the biofilm [88]. Yet no studies have been found on the application of MH in combination with PDT to eradicate bacteria. Liposomal encapsulation of plasmonic NPs and PSs as a basis for antibacterial photothermal and aPDT combination, on the other hand, has been reported. The encapsulation of gold nanorods in the liposome core and the PS curcumin in

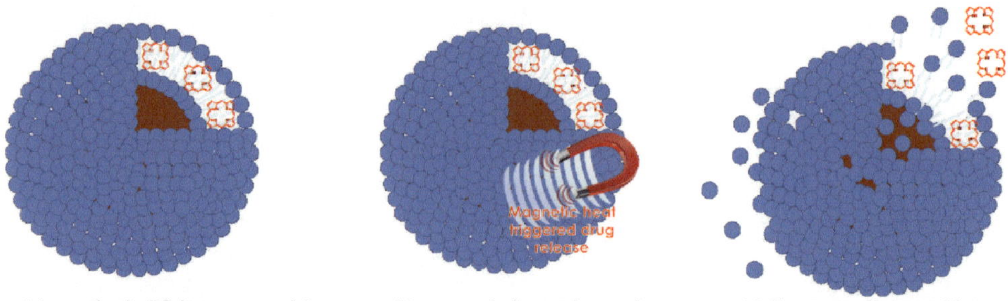

(a) porphyrin PS incorporated in the liposome bilayer

(b) magnetic heat-triggered release of the porphyrin PS

(c) liposome disassembles, releasing the porphyrin PS

Figure 9.
Magnetic hyperthermia-triggered release of the PS from the liposome bilayer.

the liposome bilayer, for example, was reported for treating recurrent acne with the combination of PTT and aPDT [89]. In this research, the curcumin PS is activated by blue light for PDT, while the gold nanorods were activated by near infrared (NIR) light for PTT, resulting in heat and ROS-based inhibition of bacterial growth.

8. Antibacterial photodynamic therapy folate targeting

A recent research report has found that folate receptor expression is significantly higher in animal tissues infected with Methicillin-resistant Staphylococcus aureus (MRSA) compared to in uninfected control tissues. The researchers exploited this finding by incorporating vancomycin in folate-decorated liposomes for folate over-expression targeting of the MRSA-infected tissue. They found that the bactericidal and biofilm inhibition effects of the folate-decorated liposomes incorporating vancomycin was higher compared to direct vancomycin application, suggesting superior MRSA targeting and delivery of the drug [90]. The targeting potential of folate functionalization was also confirmed by the superior targeting and antibacterial enhancement of the efficacy of folate-functionalized cerium NPs [91]. This has been widely exploited in PDT studies. Recently, for example, titanium dioxide NPs have been conjugated with folic acid and a phthalocyanine PS for targeted anticancer PDT [92].

The folate over-expression disease cell-targeting mechanism is illustrated in **Figure 10**. It involves the functionalization of the PS-carrier NPs with folic acid. These NPs will bind to the folate receptors followed by enhanced endocytosis by the disease cells because there is enhanced expression of the folate receptors on the disease cells. Once inside the disease cells, the NPs are induced by the disease cell internal microenvironment to release their PS cargo, thus initiating the PDT cell death pathways.

Therefore, in addition to their well-known folate over-expression-enabled cancer-cell targeting, folate-functionalized NPs incorporating PDT PSs could be used to target bacterial infection for enhanced antibacterial PDT. It is therefore quite surprising that this potential of folate targeting bacterial infection has hardly been investigated as a bacterial targeting strategy in antibacterial PDT. In this regard, the potential of the findings of the folate receptor over-expression of MRSA by Vanamala et al. [90] could be the groundbreaking research that will lead to folate-targeting applications in antibacterial PDT. Therefore, studies are required to determine the microbial infection universality of the higher folate receptor expression found in MRSA-infected tissues compared to in uninfected control tissues.

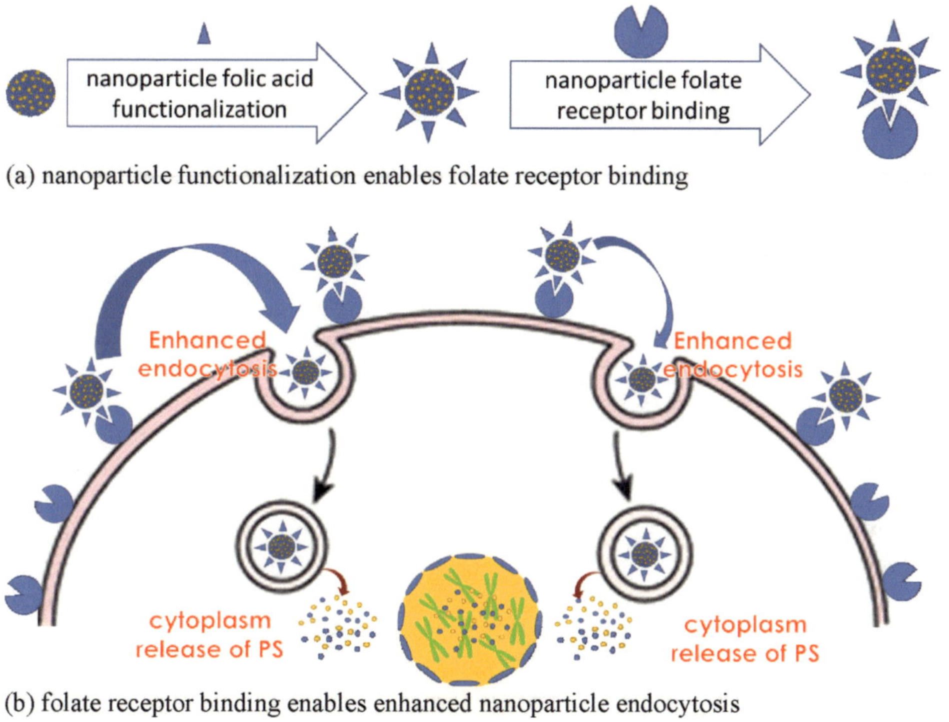

Figure 10.
The folate over-expression disease cell-targeting mechanism.

9. Nanoparticles as agents for photodynamic therapy combination therapies

Nanotechnology has taken a prime position in PDT combination therapeutics research where NPs, due to their small size and huge volume-to-surface area ratio, can absorb and otherwise load large quantities of PSs and the therapeutic agents required for the PDT combination therapy. In combining PDT with antibacterial chemotherapy, for example, NPs are loaded with antibacterial PDT PSs and antibiotic drugs. In this combination therapy, the NPs can act not only as drug and PS carriers and delivery agents but also as bacterial infection-targeting agents [93]. Nanoconjugates that are either inherently cationic or rendered cationic by virtue of cationic PS capping agents have shown selectivity for bacterial pathogens and bacterial infection. For example, polymeric chlorin-e6-incorporating nanoconjugate systems that become cationic at slightly acidic pH were reported to show pH-dependent bacterial selectivity and efficacy variation [94], while a zeolite-based inorganic NP, capped with the cationic tetravalent silicon phthalocyanine PS, showed a positive zeta potential, selectivity for bacterial infection, and enhanced efficacy [95]. However, the primary purpose of multifunctional nanoconjugate systems is to incorporate functionalities that enable the desired combination therapy into nanoconjugate systems. For example, to enable the combination of antibiotic chemotherapy with PDT, a nanoconjugate system consisting of a core of gold and a shell of silver was passivated with 4-mercaptobenzoic acid. Subsequently, the mercaptobenzoic acid shell of the gold NPs was modified by conjugation with vancomycin and loaded with the phthalocyanine PS to enable vancomycin-mediated antibiotic chemotherapy in combination with phthalocyanine-mediated PDT [96].

10. Antibacterial photodynamic therapy in combination with chemotherapy

The urgent response to the fast development of bacterial resistance to antibiotics is probably the single most compelling reason for the current rising interest in nanomaterial-mediated PDT, specifically the ability of nanomaterials to carry multiple drug and PS payloads and to deliver them only in the infection site or upon stimulation by an external source, which is the basis for combination therapies involving PDT. The reason for this is that although mechanisms of resistance against antibacterial PDT have been described, including hypoxia, the repair of DNA damage, efflux of the PS, upregulation of the heat shock protein, and inhibition of apoptosis [97], very little resistance has been observed [98–100]. Antibiotic therapy and PDT have been the subject of much interest [101–104].

Incorporating a porphyrin PS, the immunosuppressant methotrexate, and silver in one nanoconjugate may be used as an example of the nanomaterial-mediated combination of PDT with an antibiotic material [105]. While demonstrating biocompatibility and release of silver and the porphyrin PS, the nanoconjugate also showed excellent antibacterial activity in excess of that shown by the antibiotic and the porphyrin PS each acting alone. This type of combination of antibiotic silver with PDT was also shown by the eradication of *S. aureus* by a conjugate of the zinc (II) phthalocyanine PS with silver NPs [106]. Another example is the PDT treatment with amoxicillin-coated gold NPs, eradicating the embedded *P. aeruginosa* and *S. aureus*. In this example, amoxicillin was the antibiotic agent, while the nanogold acted as the PS [107]. Similarly, the biofilms of *E. coli*, *S. aureus*, and MRSA were treated with an MOF loaded with the PS methylbenzene blue and the antibiotic drug vancomycin [108]. This caused the biofilm matrix structure to collapse. The nanoconjugate was therefore able to diffuse and eradicate the bacteria. This is an example of a pH-triggered release of the antibiotic and the PS because the MOF structure disassembles upon the drop in pH as the nanoconjugate enters the bacterial and biofilm infection site and as it enters the bacterial cell into the cytoplasm.

11. Anticancer photodynamic therapy in combination with chemotherapy

Given the history of PDT and chemotherapy as anticancer therapeutic technologies, it is not surprising that the combination of PDT and anticancer chemotherapy is among the most widely reported. A study of the combination of the aluminum phthalocyanine chloride complex as the PS and doxorubicin as the chemotherapy drug agent, both encapsulated in nano-emulsions, was reported to reduce the induction of breast cancer in mice, to reduce the expression of the vascular endothelial growth factor, and to increase the expression of the apoptosis-indicating Caspase-3 protein as well as tissue death by necrosis. There was a large reduction in cancer cell proliferation [109]. The combination of PDT with anticancer chemotherapy was also studied using chlorin-e6 as the PS and cisplatin as the chemotherapy drug while ameliorating hypoxia using perfluorocarbon-mediated molecular oxygen self-supply. These three key elements were incorporated in pH- and ROS-responsive micelles made of polyethylene glycol and polyglycolic acid. This remarkable innovation showed enhanced activity against SKOV3 ovarian cancer cell lines [110]. This was attributed to the pH responsiveness, which ensured that the nano-micelles released their cargo only when they were inside the ovarian cancer cells. The perfluoro hydrocarbon-mediated oxygen encapsulation in the micelles ensured oxygenation to overcome hypoxia. **Figure 11** shows a similar nanoconjugate formed by the self-assembly of polyethylene glycol

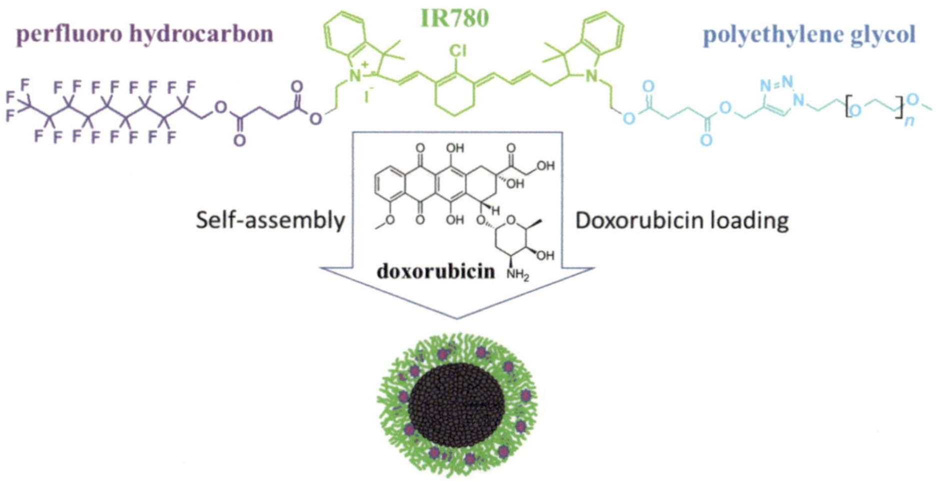

Figure 11.
Self-assembly of polyethylene glycol after conjugation with the perfluoro hydrocarbon and IR780 was loaded with doxorubicin.

after conjugation of the PS IR780 with a perfluoro hydrocarbon, along with loading with doxorubicin, achieving similar results against MCF-7 cancer cell lines [111].

12. Photodynamic therapy in combination with photothermal hyperthermia

Plasmonic NPs that possess high photothermal conversion, such as those of graphene oxide, are required to enable the combination of PDT and PTT against cancer and other diseases. For example, the conjugation of graphene oxide NPs with folic acid to enable the folate over-expression cancer cell-targeting mechanism and chlorin-e6 to enable PDT was reported to enhance the cancer cell uptake and retention and to selectively kill cancer cells *in vitro* [112]. In this study, MCF-7 cancer cell lines were studied by confocal electron microscopy. These studies revealed that the folic acid and chlorin-e6-functionalized graphene oxide NPs were localized well within the cytoplasm and not on the cell membrane. This supported the folic acid-enabled endocytosis of the graphene oxide nanoconjugate.

Gold NPs are well-known for their photoacoustic and photothermal properties, which enable light absorption and the production of heat due to their localized plasmon surface resonance [113]. Marketed as Temoporfin or Foscan over the past decade or so, the PS meso-tetrakis(3-hydroxyphenyl)chlorin is approved by the Food and Drug Administration (FDA) for PDT [114]. Therefore, a therapeutic modality that combines meso-tetrakis(3-hydroxyphenyl)chlorin-mediated PDT with gold NP-mediated PTT, based on a nanoconjugate made of NPs of gold and a capping shell of meso-tetrakis(3-hydroxyphenyl)chlorin, was reported to have a twofold synergistic enhancement of efficacy against the neuroblastoma-derived SH-SY5Y cells *in vitro* (**Figure 12**) [115]. In this study, the gold NPs were initially stabilized using mercaptopropionic acid, followed by esterification of the carboxylic acid functional groups of the mercaptopropionic acid with one of the phenolic groups of the PS, thus transforming the gold NP capping shell from carboxylic acid to a meso-tetrakis(3-hydroxyphenyl)chlorin shell. In addition to gold and graphene oxide, plasmonic NPs

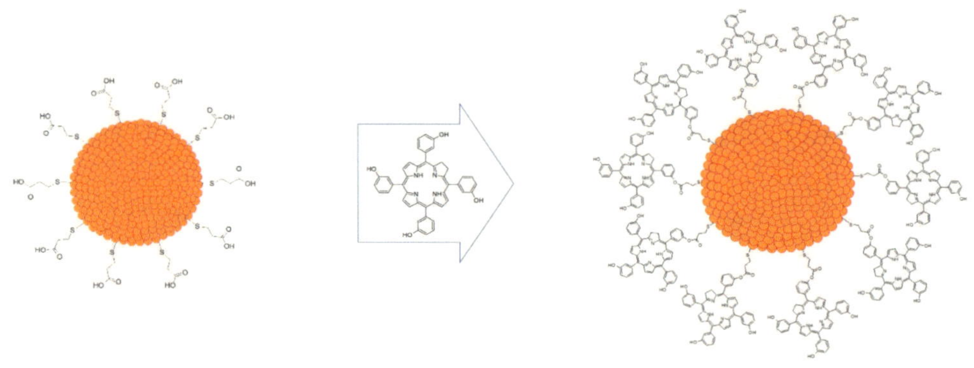

Figure 12.
A nanoconjugate made of nanoparticles of gold with a capping shell of temoporfin.

used in the combination of PDT with PTT include those of silver, silica, upconverting lanthanides, iron oxide, and several nanotubes [116].

13. Photodynamic therapy in combination with sonodynamic therapy

The exploratory clinical study of PDT combined with sonodynamic therapy (SDT) against cholangiocarcinoma in which hematoporphyrin was used as the sono-photosensitizer provides a clear indication of the state of this technology in the clinical translation trajectory [117]. This is supported by the clinical pilot studies [118] and case reports [119] that are starting to appear in the literature. Nanotechnology is ubiquitous not only as a platform for the PDT and SDT agents [120, 121], but also for other purposes of the innovation, such as hypoxia amelioration, disease targeting, imaging-guided therapy, and stimulus-responsive release. For example, reducing human serum albumin to cleave the disulfide linkages and produce free thiol groups, followed by conjugation with hemoglobin, was used to produce a nanoplatform capable of carrying oxygen. The innovative nanoplatform was used to encapsulate manganese (II) phthalocyanine as the PS and to absorb copious amount of oxygen before administration as a sono-photosensitizer that responds to light and ultrasonic activation to produce ROS, even from the insufficiently oxygenated tumor microenvironments of 4 T1 breast cancer xenografts in mice. It also enabled magnetic resonance and photoacoustic imaging [122]. Additionally, although notoriously difficult to reach by PDT, brain cancer can be reached with SDT. The PS 3-(1′-Hexyloxy)ethyl-3-devinylpyropheophorbide-a, delivered using cationic polyacrylamide NPs functionalized with tetramethyl ammonium groups, improved the kill rate of the combination of PDT and SDT relative to the individual technologies, against the U87 human glioblastoma cell line, considered to be the best model for brain cancer [123].

14. Photodynamic therapy in combination with magnetic hyperthermia

Enabled by a whole-body applicator, MH is used to treat cancerous tumors that are difficult to reach, such as tumors in the brain and those located in dark tissues, such as the liver, spleen, pancreas, and bones [124]. Recently, however, the development of a handheld MH device has been reported for more focused treatments [125].

Nanoconjugate platforms for the combination of PDT with MH need to incorporate the PS and magnetic NPs ideally in the same nanoconjugate [87]. For example, using nanoemulsions loaded with magnetic iron oxide NPs and an aluminum phthalocyanine PS, the combination of MH and PDT was reported to achieve 66% reduction in the viability of the human bone marrow mesenchymal stem cell line [126]. In addition, there are examples of multifunctional nanotechnology platforms aimed at combining PDT with MH that are selective for cancer cells. For example, a nanoconjugate of Janus nanobulets with magnetic manganese oxide heads and chlorin-e6 PS-laden mesoporous silica bodies was cloaked with the breast cancer cell membrane. Following rapid cancer cell endocytosis, the disulfide anchor of the PS onto the mesoporous silica side of the Janus nanobulet was easily cleaved upon the pH drop of the breast cancer cell internal microenvironment, thus releasing the PS [127].

15. Conclusion

The development of resistance against antibacterial drugs is the rising challenge of the century, because it was nearly one century ago that bacterial infections were dealt a lethal blow by the discovery of antibiotics, which are now facing a drastic decline in efficacy against bacterial and fungal infections, due to the development of resistance. Generally, nanomaterial-mediated targeted drug delivery is a major thrust against antibacterial drug resistance. It has been noted that the development of resistance against PDT is a difficult feat for microbial pathogens to mount. It follows therefore that nanomaterial-mediated targeted PS delivery further diminishes the likelihood of the development of resistance. The boundaries may be pushed even further by the combination of nanomaterial-mediated delivery of antibiotic drugs and PDT PS, contained in multifunctional nanoconjugate systems used in photodynamic-antibiotic chemotherapy drug combination therapies because such systems would dramatically reduce systemic release of the antibiotic chemotherapy drug and PS. It is for these reasons that the ruthenium polypyridyl complexes, which are potent generators of ROS upon photoirradiation as antibacterial PDT PSs, have been developed with the purpose of pursuing the capability of combating antibiotic resistance [128]. In the same way as combining different antibiotic drugs is effective in repurposing older drugs rendered unusable by the development of resistance, so is combining antibiotic drugs with PDT [129].

Future benefits of combination therapies include enhancement of the combating of cancer and bacterial infections. However, the current rapid expansion of the scope of these combination therapies has left a few gaps. For example, the applications of the combination of MH with PDT, which appears to have been thwarted by the requirements for onerous investments in equipment and infrastructure, could benefit in the future from the development of handheld MH devices, subject to the advances in availability of such devices [125]. Therefore, this paper is a timeous addition to the advocacy for the development of such handheld devices. Additionally, other combinations with PDT hold great promise for the future, which are still being explored in experimental research. For example, the combination of PDT with cold atmospheric pressure plasma therapy has been shown by researchers at the Universität Greifswald in Germany, to eradicate bacterial infections of common skin and wound pathogens *in vitro* [130]. This provides an initial proof of concept for what could potentially revolutionize the way in which wound infections are treated in the future, especially in the developing world where such infections unnecessarily kill many people. Therefore, more research is needed to evaluate the combination of PDT with CAP. The applicability of the folate antimicrobial targeting mechanisms across the microbial spectrum should also be determined. In conclusion, this paper has not only navigated the combination therapies that include PDT but also exposed some of the opportunities for further research and potential human benefit from it. The data show that the combinations of PDT with similarly minimally invasive technologies will further enhance the clinical translation of PDT and the development of devices that will support these combinations. Therefore, this paper encourages further research and innovation in the development of devices to be used in support of the research on combination therapies as well as clinical applications.

Glossary of acronyms

PDT	Photodynamic therapy
aPDT	Antimicrobial photodynamic therapy
MH	Magnetic hyperthermia therapy
PTT	Photothermal hyperthermia therapy
ROS	Reactive oxygen species
EPS	Extracellular polymeric substance
MOF	Metal-organic frameworks
NP	Nanoparticles
MRSA	Methicillin-resistant *S. aureus*
SDT	Sonodynamic therapy
ZIF-8	Zeolitic imidazolate metal-organic-framework-8
US	Ultrasound
PS	Photosensitizer
NIR	Near infrared
MRSA	Methicillin-resistant *Staphylococcus aureus*
FDA	Food and Drug Administration
$^1PS_0$	Singlet spin state photosensitizer in the ground state
$^1PS_1$	Singlet spin state photosensitizer in the first excited state
$^1PS_2$	Singlet spin state photosensitizer in the second excited state
$^3PS_1$	Triplet spin state photosensitizer in the first excited state
$^3O_2$	Triplet spin state molecular oxygen in the ground state
$^1O_2$	Singlet spin state molecular oxygen in the excited state

Author details

Sandile Phinda Songca
School of Chemistry and Physics, University of KwaZulu-Natal, Pietermaritzburg, South Africa

*Address all correspondence to: sandile_songca@hotmail.com

References

[1] Calixto GMF, Bernegossi J, De Freitas LM, Fontana CR, Chorilli M. Nanotechnology-based drug delivery Systems for Photodynamic Therapy of cancer: A review. Molecules. 2016;**21**:342. DOI: 10.3390/molecules21030342

[2] Songca SP, Adjei Y. Applications of antimicrobial photodynamic therapy against bacterial biofilms. International Journal of Molecular Sciences. 2022;**23**(6):3209. DOI: 10.3390/ijms23063209

[3] Huang L, Xuan Y, Koide Y, Zhiyentayev T, Tanaka M, Hamblin MR. Type I and type II mechanisms of antimicrobial photodynamic therapy: An in vitro study on gram-negative and gram-positive bacteria. Lasers in Surgery and Medicine. 2012;**44**:490-499. DOI: 10.1002/lsm.22045

[4] Muehler D, Rupp CM, Keceli S, Brochhausen C, Siegmund H, Maisch T, et al. Insights into mechanisms of antimicrobial photodynamic action toward biofilms using Phenalen-1-one derivatives as photosensitizers. Frontiers in Microbiology. 2020;**11**:589364. DOI: 10.3389/fmicb.2020.589364

[5] Kikuchi T, Mogi M, Okabe I, Okada K, Goto H, Sasaki Y, et al. Adjunctive application of antimicrobial photodynamic therapy in nonsurgical periodontal treatment: A review of literature. International Journal of Molecular Sciences. 2015;**16**:24111-24126. DOI: 10.3390/ijms161024111

[6] Guidolin K, Ding L, Chen J, Wilson BC, Zheng G. Porphyrin-lipid nanovesicles (Porphysomes) are effective photosensitizers for photodynamic therapy. Nano. 2021;**10**(12):3161-3168. DOI: 10.1515/nanoph-2021-0220

[7] Liu H, Li YY, Li X, Huang JD. Nanostructured self-assemblies of photosensitive dyes: Green and efficient theranostic approaches. Green Chemical Engineering. 2022:1-18. DOI: 10.1016/j.gce.2022.06.006 [In press, corrected proof]

[8] Chen LJ, Zhao X, Yan XP. Porphyrinic metal-organic frameworks for biological applications. Advanced Sensor and Energy Materials. 2023;**2**(1):100045. DOI: 10.1016/j.asems.2022.100045

[9] Zhuge W, Liu Y, Huang W, Zhang C, Wei L, Peng J. Conductive 2D phthalocyanine-based metal-organic framework as a photoelectrochemical sensor for N-acetyl-L-cysteine detection. Sensors and Actuators B: Chemical. 2022;**367**:132028. DOI: 10.1016/j.snb.2022.132028

[10] Abrahamse H, Hamblin MR. New photosensitizers for photodynamic therapy. Biochemical Journal. 2016;**473**(4):347-364. DOI: 10.1042/BJ20150942

[11] Curcio A, Silva A, Cabana S, Espinosa A, Baptiste B, Menguy N, et al. Iron oxide Nanoflowers @ CuS hybrids for cancer tri-therapy: Interplay of Photothermal therapy. Magnetic hyperthermia and photodynamic therapy. Theranostics. 2019;**9**(5):1288. DOI: 10.7150/thno.30238

[12] Zhang L, Zhu C, Huang R, Ding Y, Ruan C, Shen X-C. Mechanisms of reactive oxygen species generated by inorganic nanomaterials for cancer therapeutics. Frontiers in Chemistry. 2021;**9**:630969. DOI: 10.3389/fchem.2021.630969

[13] Bharathiraja S, Manivasagan P, Moorthy MS, Bui NQ, Lee KD, Oh J.

Chlorin e6 conjugated copper sulfide nanoparticles for photodynamic combined photothermal therapy. Photodiagnosis and Photodynamic Therapy. 2017;**19**:128-134. DOI: 10.1016/j.pdpdt.2017.04.005

[14] Bonnett R, Djelal BD, Nguyen A. Physical and chemical studies related to the development of m-THPC (FOSCAN®) for the photodynamic therapy (PDT) of tumours. Journal of Porphyrins and Phthalocyanines. 2001;**5**(8):652. DOI: 10.1002/jpp.377

[15] Husain D, Miller JW, Michaud N, Connolly E, Flotte TJ, Gragoudas ES. Effects of photodynamic therapy using verteporfin on experimental choroidal neovascularization and normal retina and choroid up to 7 weeks after treatment. Archives of Ophthalmology. 1996;**114**(8):978. DOI: 10.1001/archopht.1996.01100140186012

[16] Woodburn KW, Fan Q, Kessel D, Luo Y, Young SW. Photodynamic therapy of B16F10 murine melanoma with lutetium Texaphyrin. Journal of Investigative Dermatology. 1998;**110**(5):746. DOI: 10.1046/j.1523-1747.1998.00182.x

[17] Kinsella TJ, Baron ED, Colussi VC, Cooper KD, Hoppel CL, Ingalls ST, et al. Preliminary clinical and pharmacologic investigation of photodynamic therapy with the silicon phthalocyanine photosensitizer pc 4 for primary or metastatic cutaneous cancers. Frontiers in Oncology. 2011;**1**:14. DOI: 10.3389/fonc.2011.00014

[18] Kaplan MJ, Somers RG, Greenberg RH, Ackler J. Photodynamic therapy in the management of metastatic cutaneous adenocarcinomas: Case reports from phase 1/2 studies using tin ethyl etiopurpurin (SnET2). Journal of Surgical Oncology. 1998;**67**(2):121. DOI: 10.1002/(sici)1096-9098(199802)67:2<121::aid-jso9>3.0.co;2-c

[19] Mori K, Ohta M, Sano A, Yoneya S, Sonoda M, Kaneda A, et al. Potential of photodynamic therapy with a second-generation sensitizer: Mono-L-aspartyl chlorin e6. Nippon Ganka Gakkai Zasshi. 1997;**101**(2):134

[20] Dougherty TJ. New PDT studies at the Roswell Park Cancer Institute. Photodynamic Theray News. 1998;**1**(2):5

[21] Bellnier DA, Greco WR, Nava H, Loewen GM, Oseroff AR, Dougherty TJ. Mild skin photosensitivity in cancer patients following injection of Photochlor (2-[1-hexyloxyethyl]-2-devinyl pyropheophorbide-a; HPPH) for photodynamic therapy. Cancer Chemotherapy and Pharmacology. 2006;**57**(1):40. DOI: 10.1007/s00280-005-0015-6

[22] Marcus SL, Sobel RS, Golub AL, Carroll RL, Lundahl S, Shulman DG. Photodynamic therapy (PDT) and Photodiagnosis (PD) using endogenous photosensitization induced by 5-Aminolevulinic acid (ALA): Current clinical and development status. Journal of Clinical Laser Medicine & Surgery. 1996;**14**(2):59. DOI: 10.1089/clm.1996.14.59

[23] Korbelik M, Madiyalakan R, Woo T, Haddadi A. Antitumor efficacy of photodynamic therapy using novel Nanoformulations of Hypocrellin photosensitizer SL052. Photochemistry and Photobiology. 2012;**88**(1):188. DOI: 10.1111/j.1751-1097.2011.01035.x

[24] Monro S, Colón KL, Yin H, Roque J III, Konda P, Gujar S, et al. Transition metal complexes and photodynamic therapy from a tumor-Centered approach: Challenges, opportunities,

and highlights from the development of TLD1433. Chemical Reviews. 2019;**119**(2):797. DOI: 10.1021/acs.chemrev.8b00211

[25] Diaz-Diestra D, Gholipour HM, Bazian M, Thapa B, Beltran-Huarac J. Photodynamic therapeutic effect of nanostructured metal Sulfide photosensitizers on cancer treatment. Nanoscale Research Letters. 2022;**17**:33. DOI: 10.1186/s11671-022-03674-8

[26] Matlou GG, Abrahamse H. Nanoscale metal–organic frameworks as photosensitizers and nanocarriers in photodynamic therapy. Frontiers in Chemistry. 2020;**10**:971747. DOI: 10.3389/fchem.2022.971747

[27] Wang YY, Liu YC, Sun H, Guo DS. Type I photodynamic therapy by organic–inorganic hybrid materials: From strategies to applications. Coordination Chemistry Reviews. 2019;**395**:46-62. DOI: 10.1016/j.ccr.2019.05.016

[28] Rejinold NS, Choi G, Choy J-H. Bio-inorganic layered double hydroxide Nanohybrids in Photochemotherapy: A mini review. International Journal of Molecular Sciences. 2022;**23**:11862. DOI: 10.3390/ijms231911862

[29] Liang C, Zhang X, Wang Z, Wang W, Yang M, Dong X. Organic/inorganic nanohybrids rejuvenate photodynamic cancer therapy. Journal of Materials Chemistry B. 2020;**8**:4748-4763. DOI: 10.1039/d0tb00098a

[30] Bechet D, Couleaud P, Frochot C, Viriot ML, Guillemin F, Barberi-Heyob M. Nanoparticles as vehicles for delivery of photodynamic therapy agents. Trends in Biotechnology. 2008;**26**(11):612-621. DOI: 10.1016/j.tibtech.2008.07.007

[31] Qidwai A, Annu, Nabi B, Kotta, S., Narang, J.K., Baboota, S. and Ali, J. Role of nanocarriers in photodynamic therapy. Photodiagnosis and Photodynamic Therapy 2020, 30, 101782. doi:10.1016/j.pdpdt.2020.101782

[32] Knopp D, Tang D, Niessner R. Review: Bioanalytical applications of biomolecule-functionalized nanometer-sized doped silica particles. Analytica Chimica Acta. 2009;**647**(1):14-30. DOI: 10.1016/j.aca.2009.05.037

[33] Mokwena MG, Kruger CA, Mfouo-Tynga I, Abrahamse H. A review of nanoparticle photosensitizer drug delivery uptake systems for photodynamic treatment of lung cancer. Photodiagnosis and Photodynamic Therapy. 2018;**22**:147-154. DOI: 10.1016/j.pdpdt.2018.03.006

[34] Pucelik B, Sułek A, Drozd A, Stochel G, Pereira MM, Pinto SMA, et al. Enhanced cellular uptake and photodynamic effect with amphiphilic fluorinated porphyrins: The role of Sulfoester groups and the nature of reactive oxygen species. International Journal of Molecular Sciences. 2020;**21**(8):2786. DOI: 10.3390/ijms21082786

[35] Kruger CA, Abrahamse H. Utilisation of targeted nanoparticle Photosensitiser drug delivery Systems for the Enhancement of photodynamic therapy. Molecules. 2018;**23**(10):2628. DOI: 10.3390/molecules23102628

[36] Laroui H, Wilson DS, Dalmasso G, Salaita K, Murthy N, Sitaraman SV, et al. Nanomedicine, in GI. American Journal of Physiology-Gastrointestinal and Liver Physiology. 2011;**300**(3):G371-G383. DOI: 10.1152/ajpgi.00466.2010

[37] Spagnul C, Turner LC, Boyle RW. Immobilized photosensitizers for antimicrobial applications. Journal of Photochemistry and Photobiology B:

Biology. 2015;**150**:11-30. DOI: 10.1016/j.jphotobiol.2015.04.021

[38] Sanità G, Carrese B, Lamberti A. Nanoparticle surface functionalization: How to improve biocompatibility and cellular internalization. Frontiers in Molecular Biosciences. 2020;7:587012. DOI: 10.3389/fmolb.2020.587012

[39] Wan Y, Lu G, Wei W-C, Huang Y-H, Li S, Chen J-X, et al. Stable organic photosensitizer nanoparticles with absorption peak beyond 800 Nanometers and high reactive oxygen species yield for multimodality Phototheranostics. ACS Nano. 2020;**14**(8):9917-9928. DOI: 10.1021/acsnano.0c02767

[40] Alexere AMI, Sliem MA, EL-Balshy RM, Amin RM, Harith MA. Exploiting biosynthetic gold nanoparticles for improving the aqueous solubility of metal-free phthalocyanine as biocompatible PDT agent. Materials Science and Engineering: C. 2017;**76**:727-734. DOI: 10.1016/j.msec.2017.03.129

[41] Li L, Nurunnabi M, Nafiujjaman M, Jeong YY, Lee YK, Huh KM. A photosensitizer-conjugated magnetic iron oxide/gold hybrid nanoparticle as an activatable platform for photodynamic cancer therapy. Journal of Materials Chemistry B. 2014;**2**(19):2929. DOI: 10.1039/c4tb00181h

[42] Pantiushenko IV, Rudakovskaya PG, Starovoytova AV, Mikhaylovskaya AA, Abakumov MA, Kaplan MA, et al. Development of bacteriochlorophyll a-based near-infrared photosensitizers conjugated to gold nanoparticles for photodynamic therapy of cancer. Biochemistry Moscow. 2015;**80**:752-762. DOI: 10.1134/S0006297915060103

[43] Guo Y, Rogelj S, Zhang P. Rose Bengal-decorated silica nanoparticles as photosensitizers for inactivation of gram-positive bacteria. Nanotechnology. 2010;**21**(6):065102. DOI: 10.1088/0957-4484/21/6/065102

[44] Rak J, Kabesova M, Benes J, Pouckova P, Vetvicka D. Advances in liposome-encapsulated Phthalocyanines for photodynamic therapy. Life. 2023;**13**:305. DOI: 10.3390/life13020305

[45] van Nostrum CF. Polymeric micelles to deliver photosensitizers for photodynamic therapy. Advanced Drug Delivery Reviews. 2004;**56**(1):9-16. DOI: 10.1016/j.addr.2003.07.013

[46] Avci P, Erdem SS, Hamblin MR. Photodynamic therapy: One step ahead with self-assembled nanoparticles. Journal of Biomedical Nanotechnology. 2014;**10**(9):1937-1952. DOI: 10.1166/jbn.2014.1953

[47] Glatter O, Salentinig S. Inverting structures: From micelles via emulsions to internally self-assembled water and oil continuous nanocarriers. Current Opinion in Colloid and Interface Science. 2020;**49**:82-93. DOI: 10.1016/j.cocis.2020.05.003

[48] Fahmy SA, Azzazy HME, Schaefer J. Liposome photosensitizer formulations for effective cancer photodynamic therapy. Pharmaceutics. 2021;**13**(9):1345. DOI: 10.3390/pharmaceutics13091345

[49] Temizel E, Sagir T, Ayan E, Isik S, Ozturk R. Delivery of lipophilic porphyrin by liposome vehicles: Preparation and photodynamic therapy activity against cancer cell lines. Photodiagnosis and Photodynamic Therapy. 2014;**11**(4):537-545. DOI: 10.1016/j.pdpdt.2014.07.006

[50] Ghosh S, Carter KA, Lovell JF. Liposomal formulations of photosensitizers. Biomaterials. 2019;**218**:119341. DOI: 10.1016/j.biomaterials.2019.119341

[51] Nisnevitch M, Nakonechny F, Nitzan Y. Photodynamic antimicrobial chemotherapy by liposome-encapsulated water-soluble photosensitizers. Bioorganicheskaya Khimiya. 2010;**36**(3):396-402. DOI: 10.1134/s106816201003012x

[52] Zhang Y, Chang CH. Metal–organic framework thin films: Fabrication, modification, and patterning. PRO. 2020;**8**:377. DOI: 10.3390/pr8030377

[53] Das D, Yang Y, O'Brien JS, Breznan D, Nimesh S, Bernatchez S, et al. Synthesis and physicochemical characterization of mesoporous SiO2 nanoparticles. Journal of Nanomaterials. 2014;**2014**:176015. DOI: 10.1155/2014/176015

[54] Song L, Zhang J, Sun L, Xu F, Li F, Zhang H, et al. Mesoporous metal–organic frameworks: Design and applications. Energy and Environmental Science. 2012;**5**:7508-7520. DOI: 10.1039/C2EE03517K

[55] Hong SH, Choi Y. Mesoporous silica-based nanoplatforms for the delivery of photodynamic therapy agents. Journal of Pharmaceutical Investigation. 2017;**48**:3-17. DOI: 10.1007/s40005-017-0356-2

[56] Bharti C, Nagaich U, Pal AK, Gulati N. Mesoporous silica nanoparticles in target drug delivery system: A review. International Journal of Pharmaceutical Investigation. 2015;**5**(3):124-133. DOI: 10.4103/2230-973X.160844

[57] Guo Z, Wu L, Wang Y, Zhu Y, Wan G, Li R, et al. Design of Dendritic Large-Pore Mesoporous Silica Nanoparticles with controlled structure and formation mechanism in dual-Templating strategy. ACS Applied Materials and Interfaces. 2020;**12**(16):18823-18832. DOI: 10.1021/acsami.0c00596

[58] Chang M, Zhao Y, Yang Q, Liu D. Microporous metal-organic frameworks with hydrophilic and hydrophobic pores for efficient separation of CH4/N2 mixture. ACS Omega. 2019;**4**(11):14511-14516. DOI: 10.1021/acsomega.9b01740

[59] Li S, Wei Y, Wang Y, Liang H. Advances in hydrophilic metal–organic frameworks for N-linked glycopeptide enrichment. Frontiers in Chemistry. 2022;**10**:1091243. DOI: 10.3389/fchem.2022.1091243

[60] Alves SR, Calori IR, Tedesco AC. Photosensitizer-based metal-organic frameworks for highly effective photodynamic therapy. Materials Science and Engineering: C. 2021;**131**:112514. DOI: 10.1016/j.msec.2021.112514

[61] Zhang X, Wasson MC, Shayan M, Berdichevsky EK, Ricardo-Noordberg J, Singh Z, et al. A historical perspective on porphyrin-based metal–organic frameworks and their applications. Coordination Chemistry Reviews. 2021;**429**:213615. DOI: 10.1016/j.ccr.2020.213615

[62] Zhou J, Wang W, Zhang Q, Zhang Z, Guo J, Yan F. Oxygen-supplied mesoporous carbon nanoparticles for enhanced photothermal/photodynamic synergetic therapy against antibiotic-resistant bacterial infections. Chemical Science. 2022;**13**(23):6967-6981. DOI: 10.1039/d2sc01740g

[63] Guo Z, Zheng K, Tan Z, Liu Y, Zhao Z, Zhu G, et al. Overcoming drug resistance with functional mesoporous titanium dioxide nanoparticles combining targeting, drug delivery and photodynamic therapy. Journal of Materials Chemistry B. 2018;**6**:7750-7759. DOI: 10.1039/C8TB01810C

[64] Gerlini A, Colomba L, Furi L, Braccini T, Manso AS, Pammolli A, et al. The role of host and microbial factors in the pathogenesis of pneumococcal

bacteraemia arising from a single bacterial cell bottleneck. PLoS Pathogen. 2014;**10**(3):e1004026. DOI: 10.1371/journal.ppat.1004026

[65] Yeh Y-C, Huang T-H, Yang S-C, Chen C-C, Fang J-Y. Nano-based drug delivery or targeting to eradicate bacteria for infection mitigation: A review of recent advances. Frontiers in Chemistry. 2020;**8**:286. DOI: 10.3389/fchem.2020.00286

[66] Zhao Z, Ukidve A, Kim J, Mitragotri S. Targeting strategies for tissue-specific drug delivery. Cell. 2020;**181**(1):151-167. DOI: 10.1016/j.cell.2020.02.001

[67] Giamberardino A, Labib M, Hassan EM, Tetro JA, Springthorpe S, Sattar SA, et al. Ultrasensitive norovirus detection using DNA aptasensor technology. PLoS One. 2013;**8**:e79087. DOI: 10.1371/journal.pone.0079087

[68] Zhang Y, Lai BS, Juhas M. Recent advances in aptamer discovery and applications. Molecules. 2019;**24**:941. DOI: 10.3390/molecules24050941

[69] Liu M, Wang L, Lo Y, Shiu SC-C, Kinghorn AB, Tanner JA. Aptamer-enabled nanomaterials for therapeutics. Drug Targeting and Imaging. Cells. 2022;**11**:159. DOI: 10.3390/cells11010159

[70] Ommen P, Hansen L, Hansen BK, Vu-Quang H, Kjems J, Meyer RL. Aptamer-targeted drug delivery for Staphylococcus aureus biofilm. Frontiers in Cellular and Infection Microbiology. 2022;**12**:814340. DOI: 10.3389/fcimb.2022.814340

[71] Pourhajibagher M, Etemad-Moghadam S, Alaeddini M, Mousavi RS, Bahador A. DNA-aptamer-nanographene oxide as a targeted bio-theragnostic system in antimicrobial photodynamic therapy against Porphyromonas gingivalis. Scientific Reports. 2022;**12**(1):12161. DOI: 10.1038/s41598-022-16310-3

[72] Lin TX, Lai PX, Mao JY, Chu HW, Unnikrishnan B, Anand A, et al. Supramolecular aptamers on graphene oxide for efficient inhibition of thrombin activity. Frontiers in Chemistry. 2019;**7**:1-12. DOI: 10.3389/fchem.2019.00001

[73] Park JP, Shin HJ, Park SG, Oh HK, Choi CH, Park HJ, et al. Screening and development of DNA aptamers specific to several oral pathogens. Journal of Microbiology and Biotechnology. 2015;**25**(3):393-398. DOI: 10.4014/jmb.1407.07019

[74] Guo X, Wen F, Zheng N, Li S, Fauconnier ML, Wang JA. qPCR aptasensor for sensitive detection of aflatoxin M1. Analytical and Bioanalytical Chemistry. 2016;**408**:5577-5584. DOI: 10.1007/s00216-016-9656-z

[75] Pourhajibagher M, Bahador A. Aptamer decorated emodin nanoparticles-assisted delivery of dermcidin-derived peptide DCD-1L: Photoactive bio-theragnostic agent for enterococcus faecalis biofilm destruction. Photodiagnosis and Photodynamic Therapy. 2022;**39**:1572-1000. DOI: 10.1016/j.pdpdt.2022.103020

[76] Afrasiabi S, Pourhajibagher M, Raoofian R, Tabarzad M, Bahador A. Therapeutic applications of nucleic acid aptamers in microbial infections. Journal of Biomedical Science. 2020;**27**(1):6. DOI: 10.1186/s12929-019-0611-0

[77] Tra VN, Dube DH. Glycans in pathogenic bacteria-potential for targeted covalent therapeutics and imaging agents. Chemical Communications (Camb). 2014;**50**(36):4659-4673. DOI: 10.1039/c4cc00660g

[78] Guo Y, Nehlmeier I, Poole E, Sakonsinsiri C, Hondow N, Brown A, et al. Dissecting multivalent lectin-carbohydrate recognition using polyvalent multifunctional glycan-quantum dots. Journal of the American Chemical Society. 2017;**139**:11833-11844. DOI: 10.1021/jacs.7b05104

[79] Drulis-Kawa Z. Editorial: Bacterial surface Glycans as the virulence agent and the target for predators, therapy, and the immune system. Frontiers in Microbiology. 2020;**11**:624964. DOI: 10.3389/fmicb.2020.624964

[80] Verma M, Shukla AK, Acharya A. Lectin Nanoconjugates for targeted therapeutic applications. In: Acharya A, editor. Nanomaterial - Based Biomedical Applications in Molecular Imaging. Diagnostics and Therapy. Singapore: Springer; 2020. DOI: 10.1007/978-981-15-4280-0_6

[81] Masri A, Anwar A, Khan NA, Siddiqui R. The use of Nanomedicine for targeted therapy against bacterial infections. Antibiotics (Basel). 2019;**8**(4):260. DOI: 10.3390/antibiotics8040260

[82] Zhou Y, Deng W, Mo M, Luo D, Liu H, Jiang Y, et al. Stimuli-responsive Nanoplatform-assisted photodynamic therapy against bacterial infections. Frontiers in Medicine. 2021;**8**:729300. DOI: 10.3389/fmed.2021.729300

[83] Meng X, Guan J, Lai S, Fang L, Su J. pH-responsive curcumin-based nanoscale ZIF-8 combining chemophotodynamic therapy for excellent antibacterial activity. RSC Advances. 2022;**12**:10005. DOI: 10.1039/d1ra09450e

[84] Liu S, Qiao S, Li L, Qi G, Lin Y, Qiao Z, et al. Surface charge-conversion polymeric nanoparticles for photodynamic treatment of urinary tract bacterial infections. Nanotechnology. 2015;**26**:495602. DOI: 10.1088/0957-4484/26/49/495602

[85] Boccalini G, Conti L, Montis C, Bani D, Bencini A, Berti D, et al. Methylene blue-containing liposomes as new photodynamic anti-bacterial agents. Journal of Materials Chemistry B. 2017;**5**:2788-2797. DOI: 10.1039/C6TB03367A

[86] Anilkumar TS, Shalumon KT, Chen JP. Applications of magnetic liposomes in cancer therapies. Current Pharmaceutical Design. 2019;**25**(13):1490-1504. DOI: 10.2174/1389203720666190521114936

[87] Di Corato R, Béalle G, Kolosnjaj-Tabi J, Espinosa A, Clement O, Silva AK, et al. Combining magnetic hyperthermia and photodynamic therapy for tumor ablation with Photoresponsive magnetic liposomes. ACS Nano. 2015;**9**(3):2904. DOI: 10.1021/nn506949t

[88] Sun X, Dong B, Xu H, Xu S, Zhang X, Lin Y, et al. Amphiphilic Silane modified multifunctional nanoparticles for magnetically targeted photodynamic therapy. ACS Applied Materials & Interfaces. 2017;**9**:11451-11460. DOI: 10.1021/acsami.7b00647

[89] Alvi SB, Rajalakshmi PS, Jogdand A, Sanjay AY, Veeresh B, John R, et al. Iontophoresis mediated localized delivery of liposomal gold nanoparticles for photothermal and photodynamic therapy of acne. Biomaterials Science. 2021;**9**:1421-1430. DOI: 10.1039/D0BM01712D

[90] Vanamala K, Bhise K, Sanchez H, Kebriaei R, Luong D, Sau S, et al. Folate functionalized lipid nanoparticles for targeted therapy of methicillin-resistant Staphylococcus aureus. Pharmaceutics.

2021;**13**:1791. DOI: 10.3390/pharmaceutics13111791

[91] Sarkar K, Dutta K, Chatterjee A, Sarkar J, Das D, Prasad A, et al. Nanotherapeutic potential of antibacterial folic acid-functionalized nanoceria for wound-healing applications. Nanomedicine (London, England). 2023;**18**:109-123. DOI: 10.2217/nnm-2022-0233

[92] Liang X, Xie Y, Wu J, Wang J, Petković M, Stepić M, et al. Functional titanium dioxide nanoparticle conjugated with phthalocyanine and folic acid as a promising photosensitizer for targeted photodynamic therapy in vitro and in vivo. Journal of Photochemistry and Photobiology B: Biology. 2021;**215**:112122. DOI: 10.1016/j.jphotobiol.2020.112122

[93] Thomas-Moore BA, del Valle CA, Field RA, Martin MJ. Recent advances in nanoparticle-based targeting tactics for antibacterial photodynamic therapy. Photochemical and Photobiological Sciences. 2022;**21**:1111-1131. DOI: 10.1007/s43630-022-00194-3

[94] Liu P, Xu LQ, Xu G, Pranantyo D, Neoh KG, Kang ET. pH-sensitive theranostic nanoparticles for targeting bacteria with fluorescence imaging and dual-modal antimicrobial therapy. ACS Applied Nano Materials. 2018;**1**:6187-6196. DOI: 10.1021/acsanm.8b01401

[95] Strassert CA, Otter M, Albuquerque RQ, Hone A, Vida Y, Maier B, et al. Photoactive hybrid nanomaterial for targeting, labelling, and killing antibiotic-resistant bacteria. Angewandte Chemie - International Edition. 2009;**48**:7928-7931. DOI: 10.1002/anie.200902837

[96] Zhou Z, Peng S, Sui M, Chen S, Huang L, Xu H, et al. Multifunctional nanocomplex for surfaceenhanced Raman scattering imaging and near-infrared photodynamic antimicrobial therapy of vancomycin-resistant bacteria. Colloids and Surfaces B: Biointerfaces. 2018;**161**:394-402. DOI: 10.1016/j.colsurfb.2017.11.005

[97] Casas A, Di Venosa G, Hasan T, Batlle A. Mechanisms of resistance to photodynamic therapy. Current Medicinal Chemistry. 2011;**18**:2486-2515. DOI: 10.2174/092986711795843272

[98] Liu Y, Qin R, Zaat SAJ, Breukink E, Heger M. Antibacterial photodynamic therapy: Overview of a promising approach to fight antibiotic-resistant bacterial infections. Journal of Clinical and Translational Research. 2015;**1**:140-167. DOI: 10.18053/jctres.201503.002

[99] Sabino CP, Wainwright M, Ribeiro MS, Sellera FP, dos Anjos C, Baptista MS, et al. Global priority multidrug- resistant pathogens do not resist photodynamic therapy. Journal of Photochemistry and Photobiology B: Biology. 2020;**208**:111893. DOI: 10.1016/j.jphotobiol.2020.111893

[100] Al-Mutairi R, Tovmasyan A, Batinic-Haberle I, Benov L. Sublethal photodynamic treatment does not Lead to development of resistance. Frontiers in Microbiology. 2018;**9**:1699. DOI: 10.3389/fmicb.2018.01699

[101] Feng Y, Tonon CC, Ashraf S, Hasan T. Photodynamic and antibiotic therapy in combination against bacterial infections: Efficacy, determinants, mechanisms, and future perspectives. Advanced Drug Delivery Reviews. 2021;**177**:113941. DOI: 10.1016/j.addr.2021.113941

[102] Youf R, Müller M, Balasini A, Thétiot F, Müller M, Hascoët A, et al.

Antimicrobial photodynamic therapy: Latest developments with a focus on combinatory strategies. Pharmaceutics. 1995;**2021**:13. DOI: 10.3390/pharmaceutics13121995

[103] Willis JA, Cheburkanov V, Chen S, Kassab G, Soares JM, Blanco KC, et al. Antimicrobial photodynamic therapy combined with antibiotics reduces resistance and aids elimination in four resistant bacterial strains. In: Proceedings Volume 11939. Photonic Diagnosis, Monitoring, Prevention, and Treatment of Infections and Inflammatory Diseases. 2022. pp. 1193903, 1-11. DOI: 10.1117/12.2610132

[104] Pérez-Laguna V, García-Luque I, Ballesta S, Rezusta A, Gilaberte Y. Photodynamic therapy combined with antibiotics or antifungals against microorganisms that cause skin and soft tissue infections: A planktonic and biofilm approach to overcome resistances. Pharmaceuticals. 2021;**14**:603. DOI: 10.3390/ph14070603

[105] Feng G, Huang X, Jiang X, Deng T, Li Q, Li J, et al. The antibacterial effects of Supermolecular Nano-carriers by combination of silver and photodynamic therapy. Frontiers in Chemistry. 2021;**9**:666408. DOI: 10.3389/fchem.2021.666408

[106] Sen T, Nyokong T. Enhanced photodynamic inactivation of staphylococcus aureus with Schiff base substituted zinc phthalocyanines through conjugation to silver nanoparticles. Journal of Molecular Structure. 2021;**1232**:130012. DOI: 10.1016/j.molstruc.2021.130012

[107] Rocca DM, Silvero CMJ, Aiassa V, Becerra CM. Rapid and effective photodynamic treatment of biofilm infections using low doses of amoxicillin-coated gold nanoparticles. Photodiagnosis and Photodynamic Therapy. 2020;**31**:101811. DOI: 10.1016/j.pdpdt.2020.101811

[108] Chen H, Yang J, Sun L, Zhang H, Guo Y, Qu J, et al. Synergistic chemotherapy and photodynamic therapy of Endophthalmitis mediated by Zeolitic Imidazolate framework-based drug delivery systems. Small. 2019;**15**:1903880. DOI: 10.1002/smll.201903880

[109] Cabral ÁS, Leonel ECR, Candido NM, Piva HL, de Melo MT, Taboga SR, et al. Combined photodynamic therapy with chloroaluminum phthalocyanine and doxorubicin nanoemulsions in breast cancer model. Journal of Photochemistry and Photobiology B: Biology. 2021;**218**:112181. DOI: 10.1016/j.jphotobiol.2021.112181

[110] Chen Y, Zhang L, Li F, Sheng J, Xu C, Li D, et al. Combination of chemotherapy and photodynamic therapy with oxygen self-supply in the form of mutual assistance for cancer therapy. International Journal of Nanomedicine. 2021;**16**:3679-3694. DOI: 10.2147/IJN.S298146

[111] Yang G, Tian J, Chen C, Jiang D, Xue Y, Wang C, et al. An oxygen self-sufficient NIR-responsive Nanosystem for enhanced PDT and chemotherapy against hypoxic Tumors. Chemical Science. 2019;**10**:5766-5772

[112] Guo S, Song Z, Ji D-K, Reina G, Fauny J-D, Nishina Y, et al. Combined Photothermal and photodynamic therapy for cancer treatment using a multifunctional graphene oxide. Pharmaceutics. 2022;**14**:1365. DOI: 10.3390/pharmaceutics14071365

[113] Vines JB, Yoon J-H, Ryu N-E, Lim D-J, Park H. Gold nanoparticles for Photothermal cancer therapy. Frontiers in Chemistry. 2019;**7**:167. DOI: 10.3389/fchem.2019.00167

[114] Baskaran R, Lee J, Yang SG. Clinical development of photodynamic agents and therapeutic applications. Biomaterials Research. 2018;**22**:25. DOI: 10.1186/s40824-018-0140-z

[115] Varon E, Blumrosen G, Sinvani M, Haimov E, Polani S, Natan M, et al. An engineered Nanocomplex with photodynamic and Photothermal synergistic properties for cancer treatment. International Journal of Molecular Sciences. 2022;**23**:2286. DOI: 10.3390/ijms23042286

[116] Shibu ES, Hamada M, Murase N, Biju V. Nanomaterials formulations for photothermal and photodynamic therapy of cancer. Journal of Photochemistry and Photobiology C: Photochemistry Reviews. 2013;**15**:53-72. DOI: 10.1016/j.jphotochemrev.2012.09.004

[117] Jiang X. An exploratory clinical study of photodynamic therapy combined with sonodynamic therapy against cholangiocarcinoma. US National Library of Medicine. 2022. ClinicalTrials.gov Identifier NCT05580328

[118] Zhang W, Li K, Lu J, Peng Z, Wang X, Li Q, et al. Sonodynamic and photodynamic therapy in breast cancer - a pilot study. International Journal of Complementary and Alternative Medicine. 2017;**9**(5):00313. DOI: 10.15406/ijcam.2017.09.00313

[119] Zhang W, Li K, Li LQ, Wang X, Bo X, Xu H. Sonodynamic and photodynamic therapy in advanced pancreas carcinoma - a case report. Online Journal of Complementary & Alternative Medicine. 2020;**3**:2. DOI: 10.33552/OJCAM.2020.03.000559

[120] Li R, Chen Z, Dai Z, Yu Y. Nanotechnology assisted photo- and sonodynamic therapy for overcoming drug resistance. Cancer Biology & Medicine. 2021;**18**:388-400. DOI: 10.20892/j.issn.2095-3941.2020.0328

[121] Liao S, Cai M, Zhu R, Fu T, Du Y, Kong J, et al. Antitumor effect of photodynamic therapy/Sonodynamic therapy/Sono-photodynamic therapy of Chlorin e6 and other applications. Molecular Pharmaceutics. 2023;**20**(2):875-885. DOI: 10.1021/acs.molpharmaceut.2c00824

[122] Yin T, Yin J, Ran H, Ren Y, Lu C, Liu L, et al. Hypoxia-alleviated sonodynamic therapy based on a hybrid protein oxygen carrier to enhance tumor inhibition. Biomaterials Science. 2022;**10**:294-305. DOI: 10.1039/D1BM01710A

[123] Borah BM, Cacaccio J, Durrani FA, Bshara W, Turowski SG, Spernyak JA, et al. Sonodynamic therapy in combination with photodynamic therapy shows enhanced long-term cure of brain tumor. Scientific Reports. 2020;**10**(1):21791. DOI: 10.1038/s41598-020-78153-0

[124] Maier-Hauff K, Rothe R, Scholz R, Gneveckow U, Wust P, Thiesen B, et al. Intracranial thermotherapy using magnetic nanoparticles combined with external beam radiotherapy: Results of a feasibility study on patients with glioblastoma Multiforme. Journal of Neuro-Oncology. 2007;**81**:53-60. DOI: 10.1007/s11060-006-9195-0

[125] Castro-Torres JL, Méndez J, Torres-Lugo M, Juan E. Development of handheld induction heaters for magnetic fluid hyperthermia applications and in-vitro evaluation on ovarian and prostate cancer cell lines. Biomedical Physics & Engineering Express. 2023;**9**:035010. DOI: 10.1088/2057-1976/acbeaf

[126] de Paula LB, Primo FL, Pinto MR, Morais PC, Tedesco AC. Combination of

hyperthermia and photodynamic therapy on mesenchymal stem cell line treated with chloroaluminum phthalocyanine magnetic-nanoemulsion. Journal of Magnetism and Magnetic Materials. 2015;**380**:372-376. DOI: 10.1016/j.jmmm.2014.10.098

[127] Wang Z, Zhang F, Shao D, Chang Z, Wang L, Hu H, et al. Immunotherapy: Janus Nanobullets combine photodynamic therapy and magnetic hyperthermia to potentiate synergetic anti-metastatic immunotherapy (Adv. Sci. 22/2019). Advanced Science (Weinh). 2019;**6**(22):1970136. DOI: 10.1002/advs.201970136

[128] Jain A, Garrett NT, Malone ZP. Ruthenium-based photoactive Metalloantibiotics. Photochemistry and Photobiology. 2022;**98**(1):6-16. DOI: 10.1111/php.13435

[129] Sharma J, Sharma D, Singh A, Sunita K. Colistin resistance and management of drug resistant infections. Canadian Journal of Infectious Diseases and Medical Microbiology. 2022;**2022**:4315030. DOI: 10.1155/2022/4315030

[130] Aly, F. Synergistic Antimicrobial Effects of Cold Atmospheric Pressure Plasma and Photodynamic Therapy against Common Skin and Wound Pathogens In-Vitro. Ph.D. Thesis, Universität Greifswald, Greifswald, Germany, 2021. Available online: https://epub.ub.uni-greifswald.de/frontdoor/index/index/docId/5680 [Accessed: January 16, 2022]